BIOETHICS WITH LIBERTY AND JUSTICE

BIOETHICS WITH LIBERTY AND JUSTICE

Themes in the Work of Joseph M. Boyle

Edited by
CHRISTOPHER TOLLEFSEN
University of South Carolina, Columbia, SC, USA

Springer

Editor
Prof. Christopher Tollefsen
Department of Philosophy
University of South Carolina
Columbia, South Carolina 29208
USA
Christopher.Tollefsen@gmail.com

ISSN 0376-7418
ISBN 978-90-481-9790-3 e-ISBN 978-90-481-9791-0
DOI 10.1007/978-90-481-9791-0
Springer Dordrecht Heidelberg London New York

Printed on acid-free paper

Springer is part of Springer Science+Business Media (www.springer.com)

Preface

Bioethics with Liberty and Justice: Themes in the Work of Joseph M. Boyle brings together the work of a number of philosophers, theologians, and bioethicists to pay tribute to Joseph Boyle. Over the past forty years, Boyle has contributed to the development of Catholic thought on bioethical issues across a number of areas, many of which are represented in this volume, including: ethical controversies at the beginning and end of life; the role of double effect in medical decision making; the right to health care; the development of a global bioethics; and the application of natural law thinking in medical ethics casuistry.

In addition to this professional role, Boyle has also been a colleague, mentor, and friend to many of this volume's contributors. It is with gratitude for these many roles he has played in our lives, admiration for his philosophical acumen, and deep affection for Boyle himself, that these essays are offered. We all look forward to many more years of philosophical and personal engagement with Joe on these and other issues.

As the editor of *Bioethics with Liberty and Justice*, I would like to add a special word of thanks to the various authors here assembled for their fine work; it has been a pleasure working with all of them. Thanks are due also to H. Tristram Engelhardt, Jr., for his stewardship of the *Philosophy and Medicine* series, and to Lisa Rasmussen, Associate Editor of that series, for her tremendous assistance, with this, and other volumes in the *Catholic Studies in Bioethics* series.

Columbia, South Carolina — Christopher Tollefsen

Contents

Contributors

Joseph M. Boyle University of Toronto, Toronto, ON, Canada, jboyle@chass.utoronto.ca

E. Christian Brugger Saint John Theological Seminary, Denver, CO, USA, Christian.Brugger@archden.org

Timothy Chappell Department of Philosophy, Ethics Center, The Open University, Milton Keynes, MK7 6BJ, UK, T.Chappell@open.ac.uk

Mark J. Cherry St. Edward's University, Austin, TX, USA, markc@stedwards.edu

Robert P. George Princeton University, Princeton, NJ, USA, rgeorge@princeton.edu

Germain Grisez Mount St. Mary's University, Emmitsburg, MD, USA, grisez@msmary.edu

Ana S. Iltis Center for Bioethics, Health and Society, Wake Forest University, Winston-Salem, NC, USA, iltisas@wfu.edu

Patrick Lee Franciscan University of Steubenville, Steubenville, OH, USA, plee512@gmail.com

Donald Marquis University of Kansas, Lawrence, KS, USA, dmarquis@ku.edu

R. Mary Hayden Lemmons University of St. Thomas, St. Paul, MN, USA, rmlemmons@stthomas.edu

Andrew Lustig Davidson College, Davidson, NC, USA, anlustig@davidson.edu

Peter F. Ryan, S.J. Mount St. Mary's Seminary, Emmitsburg, MD, USA, pryan@msmary.edu

Janet Smith Sacred Heart Major Seminary, Detroit, MI, USA, profjanetsmith@comcast.net

Part I
The Substantial Identity Thesis

Chapter 1
Why Abortion is Seriously Wrong: Two Views

Donald Marquis

The purpose of this essay is to compare the substantial identity argument for the wrongness of abortion to the future of value argument for its wrongness. Both arguments take for granted the standard moral judgment that it is wrong intentionally to end the lives of innocent post-natal children and adults. According to the substantial identity account such killings are wrong because an individual has the right to life in virtue of being an innocent human being and post-natal children and innocent adults are innocent human beings. It follows that, since fetuses are innocent human beings it is wrong intentionally to end their lives. According to the future of value account such killings are wrong because depriving a human individual of all of the goods of life that she would have experienced had she not been killed makes such killing wrong and innocent post-natal children and adults have such futures of value. It follows that, since fetuses have futures (very much!) like ours ending their lives is presumptively seriously wrong (Marquis, 1989).

The future of value argument and the substantial identity argument each have two parts. (1) They each offer an account of what makes killing presumptively seriously wrong in the cases of all those individuals whom, we all agree, it is seriously wrong to kill. (2) They each point out that their account implies that it is presumptively seriously wrong to end the life of a fetus.

1.1 The Future of Value Account

Although I may, with some justification, be accused of bias in this matter, I believe that the future of value account of the wrongness of killing has many virtues.

D. Marquis (✉)
University of Kansas, Lawrence, KS, USA
e-mail: dmarquis@ku.edu

This essay was begun while I was Laurance S. Rockefeller Visiting Professor for Distinguished Teaching at Princeton University and was motivated by discussions with Robert George and Patrick Lee while I was there. I am deeply indebted to Princeton University for the opportunity provided by that appointment and to Pat and Robby for prodding me intellectually.

C. Tollefsen (ed.), *Bioethics with Liberty and Justice*, Philosophy and Medicine 110,
DOI 10.1007/978-90-481-9791-0_1,

1. The future of value view is part of a more general understanding of what makes one's premature death, no matter what the cause, one's misfortune. Persons facing premature death from cancer or AIDS believe that not being able to experience the good things contained in the future lives they had thought they would live is their great misfortune. The future of value account is based on what such people actually believe.
2. The future of value account is based on what has value. It is based on what, in the victim's future, she would value. Therefore, it is superior to those accounts of the wrongness of killing based on some property not essentially connected with value.
3. The future of value view is a victim-centered view. The fundamental wrong of killing is located in something about the victim of the killing, rather than, say, a loss to society, or a loss to the world as a whole.
4. The future of value view locates the wrong of killing just where it should be in another respect. It locates the wrong of killing in the prospective future of the victim. When Fred kills Joe, Fred does not alter Joe's past. Joe's past is already completed. It cannot be changed. Furthermore, when Fred kills Joe, Fred does not really alter Joe's present. The present is instantaneous; it divides the past and the present. Because killing should make a difference to the victim, it must concern the victim's prospective future.
5. The future of value view is plausible because it fits killing into the category of harm. As a rule a shorter life contains fewer of life's goods than a longer life. Killing someone harms her by making her life shorter, and therefore (typically) worse, than it otherwise would have been. Therefore, because killing someone makes her worse off, killing her harms her.
6. The future of value argument has the virtue of not basing the wrong of killing an individual at some time on that individual's positive attitude toward her future life at that time. Positive attitude accounts of the wrong of killing are popular in the philosophical literature. Versions of such accounts refer to one's hope for one's future (Brown, 2000), or to one's desire to live (Tooley, 1972; Singer, 1979), or to one's plans and projects concerning one's future life (DeGrazia, 2005), or to one's caring about one's future (Reiman, 1999), or to one's valuing one's future (Harris, 1999), or to one's taking an interest in one's future (Dworkin, 1993; Steinbock, 1992). Fetuses lack positive attitudes toward their own futures. Therefore, positive attitude theories underwrite the moral permissibility of abortion. Furthermore, such accounts have the virtue of building the victim's values into the account.

Nevertheless, positive attitude accounts of the wrongness of killing are unsound. They do not account for the wrongness of killing human adults who lack a positive attitude toward their futures. For example, they do not account for the wrongness of killing those who suffer from severe depression, or those who ask you to kill them because they want to sacrifice their lives to the gods or to those who have been given some drug that causes them to want to die.[1] The future of value view is not subject to such counterexamples.

1.2 The Substantial Identity Account

Most people who believe that abortions are presumptively seriously wrong base their views on the substantial identity account of the wrongness of killing rather than the future of value account. The substantial identity account seems to be a fundamental part of our present moral consensus and the Western moral tradition. If we ask why the Holocaust was deeply evil, for example, virtually everyone would say that it was evil because those six million Jews were human beings and the right to life is a basic human right, a right that one has in virtue of one's humanity. When we talk about the wrongness of slavery or the wrongness of female genital mutilation or the wrongness of genocide, we talk about the violation of basic human rights. We think of human rights as rights that one has in virtue of being a human being. Presumably the right to life is the most basic of those rights.

An examination of the abortion controversy provides further evidence that the substantial identity view is part of almost everyone's moral conceptual framework. Many people on both sides of the controversy frame the abortion issue in terms of the question of when life begins.[2] Opponents of abortion choice often claim that life begins at conception, and, because at conception a new human life is created, abortion is wrong. Those who defend abortion choice disagree. They often say we don't know when (during pregnancy) life begins. Presumably, they are thinking that because we don't know when a fetus becomes a new human life, we are in no position to say that an abortion is ending a human life and therefore, do not have moral grounds for arguing that a woman's choice to abort is wrong. Apparently many opponents and also many defenders of abortion choice take the substantial identity view for granted.

The more philosophically oriented participants in the abortion debate also share a common concern. They are both concerned with arbitrariness. Joseph Boyle's views are typical of those who defend the substantial identity view. He claims that basing the wrongness of killing us on a criterion narrower than our membership in the species of human beings is "arbitrary, sectarian, and in many cases self-serving" (Boyle, 1979, 66). Boyle argues for this arbitrariness thesis in some detail.

Opponents of the substantial identity view also appeal to arbitrariness. Peter Singer has fleshed out the appeal as well as anyone. That we are human beings, and that fetuses are human beings, are biological facts. As a general rule, biological facts aren't morally significant. That we have one Y chromosome does not give us more rights than those who do not have two X chromosomes. That we are Caucasian does not give us more rights than those who are not. Accordingly, why should the biological fact that we are human have any great moral significance (Singer, 1979, 117)? Singer has labeled the view that the biological property of being a human being makes, without further argument, a moral difference "speciesism". Singer's point is that the connection between the biological and the moral of which substantial identity theorists are so fond is arbitrary.

What conclusions may we draw from these charges of arbitrariness? Arbitrariness arguments are arguments for rejecting an opposing view. They are essentially negative arguments. They do not give us reasons for accepting the view we are urged to accept. They do not give us reasons why killing is wrong in the cases of beings who are supposed to fall under the scope of the maxim "Killing is wrong". Singer has offered positive arguments that many people find plausible and principled for his account of why killing standard adults and post-natal children is wrong. I believe such arguments are flawed. Because this essay is not about arguments in defense of abortion choice, I shall not discuss them here.

Can substantial identity theorists offer positive arguments for their view? Boyle has argued that the use of species membership as a criterion the right to life is "reasonable because of the common meaning of "person", the facility of using species membership as a criterion for personhood, the arbitrariness in the use of other criteria, and the demand of justice which requires that those who might very well be persons not be arbitrarily denied consideration of this claim" (Boyle, 1979, 70). This positive defense of the substantial identity view is not persuasive. Boyle is flatly wrong about the common meaning of "person". Singerians—and there are many of them—find Boyle's meaning unintuitive. Boyle's last two reasons appeal again to arbitrariness, which is not a positive appeal at all, but an appeal to the insufficiency of other criteria. Boyle's appeal to the facility of treating species membership as a criterion for personhood is equally applicable to birth as a criterion for the right to life. Plainly Boyle's positive defense of the substantial identity view is unsatisfactory.

If Boyle's positive defense of the substantial identity view were the only positive defense of it available, then the substantial identity view should be cast to the flames, not because Singer's view is clearly superior, but because there is no good reason for favoring the substantial identity view over Singer's view, and if this is the case (and the future of value view is not in the picture), then why not choose reproductive freedom? After all, freedom is clearly a good.

However, we are not forced to accept this conclusion. Francis Beckwith, Patrick Lee and Robert George have recently offered an interesting positive defense of the substantial identity view. Lee and George have set out the substantial identity view as follows (I have made some friendly alterations to it; Lee and George, 2008, 134; Beckwith, 2007, 161).

1. You and I have the right to life.
2. We have the right to life in virtue of being intrinsically valuable.
3. We are intrinsically valuable in virtue of what we are essentially.
4. We are essentially human physical organisms.
5. Therefore, we have the right to life in virtue of being human physical organisms.

This argument is plainly valid. If it is sound, then, since fetuses are human physical organisms, a compelling defense of a fetal right to life is immediately apparent. Notice also that (1), (2), and (4) are apparently uncontroversial, although a few defenders of abortion choice reject (4).[3] I shall not now enter the muddy waters that swirl around (4). (3) does the real work in this argument for most of us. A defense of (3) requires positive arguments for it. Beckwith, Lee and George have offered such arguments.

The meaning of (3) is worth a comment. Talk about what we are essentially is talk about our essential properties. A property of a thing is essential if and only if it is essential, that is necessary, to a thing's *existence*. In the absence of that property the thing longer exists. It follows that an essential property of a thing is a property it has as long as it exists. Those who defend the substantial identity view wish to argue that what gives us the right to life is an essential, not an accidental property, of us. For the purposes of this discussion, an accidental property of us is any property of us that is not essential.

The substantial identity account's recent defenders have appealed to assumptions that seem to be at least first cousins of a view that Singer has defended. Singer has argued that there is special value in the life of a person. Suppose one surveys all of the individuals in the universe. Suppose one asks: what precisely is it that makes us morally special so that individuals like us have the right to life and others do not? Singer's answer is: We are persons (Singer, 1979, 78). Singer means by "person" a rational and self-conscious being' (Singer, 1979, 76).

Francis Beckwith's recent defense of the substantial identity account of the wrongness of killing rejects the charge of speciesism. According to Beckwith "The pro-life position is based on the *personal nature* of human beings" (Beckwith, 2007, 161). He says that if we discovered another species that had a nature like ours, then members of that species would also have the right to life (Beckwith, 2007, 162). Patrick Lee and Robert George ask how the line between those who have the right to life and those who do not can be drawn in a non-arbitrary way. "We will argue that the criterion is having a rational nature..." (Lee and George, 2008, 82). They explicitly admit that there may be other species with a rational nature.

The Lee, George, and Beckwith defense of the substantial identity view can be made perspicuous in terms of the following syllogism in Barbara:

1. All individuals who are rational beings are individuals who have the right to life.
2. All human beings are individuals who are rational beings.
3. Therefore, all human beings are individuals who have the right to life.

Lee, George and Beckwith need to affirm (2), because without (2) their inference from (1) to (3) would be invalid. Because this syllogism is so central to their recent defense of the substantial identity account and because I shall refer to it throughout this essay, I shall name it. Let us call it, after its authors, "BLG".

BLG is attractive. The truth of the major premise *under some interpretation* is not in doubt and I shall take its truth for granted in this essay. Singer, Lee, George and Beckwith call anyone who satisfies the conditions of the major premise "persons". I shall use that terminology. Singer rejects the conclusion of BLG because he believes that the minor premise is false. In particular, he claims that, although fetal human beings are (trivially) human beings, no fetus is yet a rational being. Therefore, the issue between Singer and the substantial identity theorists concerns the criteria for attributing "is a rational being" to an individual; it concerns the correct understanding, but not the truth, of the major premise.

How should BLG's major premise be understood? Lee has set out the alternatives clearly (Lee, 2004, 252–253). The first alternative is that a necessary condition for attributing "is a rational being" to an individual at a time is in terms of the actual exhibition of signs of rationality and self-awareness at that same time. This possibility can be ruled out quickly because if it were true, then an individual would not be a rational being when asleep. Accordingly, we must understand "is a rational being" in terms of having the *capacity* to exhibit rational agency. (I treat "is a rational being" and "has the capacity to exhibit rational agency" as equivalent.) A second candidate interpretation is that an individual has the capacity to exhibit rational agency if and only if this capacity is immediately exercisable. A third candidate interpretation is that an individual has the capacity to exhibit rational behavior if and only if she has the basic, natural capacity to exhibit signs of rational agency. Singer interprets the major premise in the second way. Beckwith, Lee, and George interpret the major premise in the third way. In order to discuss which interpretation is correct when used in the major premise of BLG, we need some reasonably clear account of the difference between an immediately exercisable capacity and a basic natural capacity.

Here is the best I can do. Fred, who is a regular chess player, has the immediately exercisable capacity to play chess even when he is not playing chess and even when he is anaesthetized for surgery. Bill, who does not understand the rules of chess at all and who has never played the game, is bright enough to learn the game. He has the basic, natural capacity to play chess. He lacks the immediately exercisable capacity to play chess. He could not play the game without acquiring some chess skills. We acquire the skills or the physical or mental abilities to perform various activities at various times throughout our lives. Before we acquired those skills, we lacked the immediately exercisable capacity to, for example, play the viola, or do proofs in symbolic logic or speak Chinese or play tennis or juggle, but if it is possible for us to acquire the relevant immediately exercisable capacity, then we have the basic natural capacity to do any of those things.

One caveat is necessary. One should not assume that one can have the basic natural capacity to do X *at some time* only if one could, or could have, *at that same time* have acquired the immediately exercisable capacity to do X. Imagine that we could have done some very sophisticated sort of brain scan on Albert Einstein when he was an infant. We might be then able to judge that Einstein had, as an infant, the basic natural capacity to be a great physicist, even though he could not have developed the immediately exercisable capacity to be a great physicist when he was an infant.

Should we, with Singer, understand the major premise of BLG in terms of the immediately exercisable capacity for rationality and self awareness, or should we, with Beckwith, Lee, and George, understand the major premise in terms of the basic natural capacity for rationality and self awareness? The two most important arguments that Lee and George offer against the former alternative and for the latter we might call "the continuum argument" and "the equality argument."[4]

Here are Lee and George on the continuum argument:

> . . .the difference between these two types of capacity is merely a difference between stages along a continuum . . . The capacities for reasoning, deliberating, and making free choices are gradually developed. . . .But the difference between a being that deserves full moral respect and a being that does not . . . cannot consist only in the fact that, while both have some feature, one has more of it than the other. A mere *quantitative* difference . . . cannot be itself be a justificatory basis for treating different entities in *radically* different ways (Lee and George, 2008, 137).

What precisely is this kind of capacity of which one has more as one develops and, conversely, of which one has less the shorter the time that has elapsed since one's conception? Can this be the basic natural capacity to exhibit rational behavior? I have had that basic natural capacity as long as I have been alive. *That capacity* does not vary in degree over the time of its development (although its behavioral manifestations may increase over time). Lee, George, and Beckwith would agree. Therefore, the feature to which Lee and George are referring is not our basic natural capacity to be rational. Can this feature be the immediately exercisable capacity to be rational and self-aware? Lee and George are thinking primarily about fetuses. Nevertheless, no fetus has at *any* time during its development *any* immediately exercisable capacity to be rational and self-aware. Lee, George (and, of course, Singer) would agree. Lee and George say as much: "A six-week old baby lacks the *immediately (or nearly immediately) exercisable* capacity to perform characteristically human mental functions" (Lee and George, 2008, 136). Therefore, it is reasonable to infer that they believe that no fetus has the immediately exercisable capacity to perform characteristically human mental functions. The feature that Lee and George presumably are talking about that develops so that there is more of it as the gestational age of a fetus increases does not exist. Of course, the immediately exercisable capacity to perform characteristically human mental functions develops by degrees *after* we are born, but that poses no difficulty for Singer's view. On his view once that capacity begins to develop we are, in virtue of that fact, persons. On his view, drawing the line between individuals who are persons and individuals who are not does not involve drawing a line at some arbitrary time during the development of the immediately exercisable capacity for rationality. What is correct about the Lee-George continuum point is not actually an objection to Singer's understanding of the capacity to exhibit rational behavior at all. One can realize that the continuum argument fails once one tries to get clear just what it is about.

Another reason for suspicion concerning the Lee-George continuum argument is that it does not apply to capacities in general. I do not now have the immediately exercisable capacity *at all* to compete in the pole vault, to speak Chinese, or to play the violin. I am hardly alone in these respects. The corresponding basic natural capacities may vary throughout our lives. For example, I probably had the basic natural capacity to learn Chinese when I was 5 years old. If I reach the age of 85, then I probably will no longer have the basic natural capacity to learn Chinese. However, this variation is not something to which Lee and George can appeal in their argument. They want to argue that our rational nature, that is, our basic natural capacity to be rational agents, is something that we have throughout our lives from conception to death. Denial of that thesis is incompatible with their substantial identity view.

Here is the Lee-George presentation of the equality argument:

> But the acquired qualities that could be proposed as criteria for personhood come in varying and continuous degrees: there is an infinite number of degrees of the *development of* the basic natural capacities for self-consciousness, intelligence, or rationality. So, if human beings were worthy of full moral respect (as subjects of rights) only because of such qualities, and not in virtue of the kind of being they are, then, since such qualities come in varying degrees, no account could be given of why basic rights are not possessed by human being in varying degrees. The proposition that all human beings are created equal would be relegated to the status of a superstition (Lee and George, 2008, 138).

This Lee-George argument apparently comes to this. If one's intrinsic value is based on being a person and being a person is based on the immediately exercisable capacity to think and behave rationally, as Peter Singer understands it, then since some people have a greater immediately exercisable capacity to exhibit rational behavior than others, those same people would have more intrinsic value than others. If some people have more intrinsic value than others, then there is no basis for the view that the right to life is a right possessed equally by all who have intrinsic value. Therefore, the right to life cannot be based upon rationality understood in terms of an immediately exercisable capacity.

Lee and George are correct: the acquired qualities Singer proposed as sufficient conditions for personhood do come in varying degrees. The trouble is that the basic natural capacities that Lee and George propose as sufficient conditions for personhood also come in varying degrees.[5] After all, some people are naturally dim-witted and some are naturally very smart. Therefore, some people have a greater natural capacity for rational agency than others. Accordingly, if the Lee-George equality argument against Singer's view were successful, then a similar equality argument would show that the Lee-George view is unsound.

Lee and George are apparently aware of this apparent difficulty with their view (Lee and George, 2008, 138n.). Here is how they deal with it: "However, the conclusion of Lee's argument was not that the criterion for the right to life is natural capacities, but that it is *being a certain type of substance*" (Lee and George, 2008, 139n.).

Will this do? Let us leave aside certain fanciful genetic engineering scenarios. It is plainly true that any given individual either is a human being or she is not a human being. A naturally dim-witted human being is no less a human being than Albert Einstein. Therefore, the substantial identity view is not subject to the equality objection. Lee and George are quite correct about this.

A difficulty remains. Lee and George claimed to be arguing for a particular interpretation of the major premise of BLG. To spell it out, they were arguing for:

1. All individuals who have the basic natural capacity for rationality are individuals having the right to life.

However, in response to a criticism of their equality argument for (1) they claim (in essence) to have been arguing for the *conclusion* of BLG which is:

3. All human beings are individuals having the right to life.

They are quite correct that (3) is not vulnerable to that criticism, but that conclusion is not what they were arguing for.

It is easy to overlook this equivocation. If one is attached to the Aristotelian philosophical tradition (which is a good tradition to admire) and one, following Aristotle, thinks that human beings are defined as rational animals, then one will not distinguish between (1) and (3). Nevertheless, Lee and George must distinguish between (1) and (3), for they explicitly claim that it is possible that there are rational agents who are not humans.

Here is the Lee-George attempt to escape this difficulty:

> However, the criterion for full moral worth is having a nature that entails the capacity (whether existing in root form or developed to the point at which it is immediately exercisable) for conceptual thought and free choice—and not *the development* of that natural basic capacity to some degree or other (and to what degree would necessarily be an arbitrary matter) (Lee and George, 2008, 93).

This misses the point. The objection to their equality argument is not that *the development* of the basic natural capacity can be a matter of degree (which it plainly is), but that *the basic natural capacity itself, whether developed to any particular degree or not* comes in degrees.

The Lee-George equality argument fails for another reason as well. The alleged equality difficulty with Singer's view—or with their own—could be avoided by understanding the capacity for rationality (whichever kind of capacity one has in mind) as a threshold criterion for the right to life. Threshold criteria are common. Although some people are better drivers than others, all who meet threshold criteria for a driver's license have an equal right to drive. Although some very good bridge players are better than other very good bridge players, the bridge tournament can be open to all life masters on an equal basis. Although some students may have stronger academic resumes than others, and although having a strong academic resume is the most important criterion for being admitted to Princeton University, all of the students admitted to Princeton University have certain equal rights in virtue of being incoming freshmen. Therefore, it is not inconsistent to consider the capacity for rationality the criterion for the right to life, and to realize that rationality comes in degrees, and nevertheless, to hold the view that everyone who meets a certain rationality threshold has the right to life and has it equally (whether one adopts the Lee-George view or the Singer view). Indeed, this seems to be the view that Lee, George, and Singer really hold.[6]

Consider a different equality argument. Let us, for a moment, set aside the considerations concerning kinds of capacities, and therefore BLG, altogether. We might take Lee and George to be arguing that *only* the substantial identity view, that is, only the view that the humanity of each of us is the basis for our intrinsic value, is compatible with the claim that human rights, including importantly the right to life, are rights we each have equally. In other words, the moral fact that we all have the right to life equally presupposes that we have that right in virtue of being human beings.[7] Is this the basis for a successful argument?

One reason that this line of argument is not, as it stands, successful, is that, in order to be successful, this line of argument would have to rule out the adequacy of threshold views. But Lee, George, and Beckwith fail to do this. Another reason is that it is certainly not clear that no philosophical account of the basis of human equality other than the substantial identity view is adequate. Perhaps there are hypothetical contract accounts that will do the job. Perhaps the Kantian moral principle that one should act only on that maxim that one can at the same time will to be a universal law will provide a basis for human equality. Lee, George and Beckwith do not consider these alternatives. However, in the absence of these considerations, this version of the equality argument is not compelling.

Let us appraise where the analysis so far has taken us. The substantial identity view is basically the view that we have the right to life in virtue of being human beings. This view can be defended in various ways. It might be defended on religious grounds. However, theological considerations are outside the scope of this discussion and are, at any rate, clearly an insufficient basis for the doctrine that abortion is wrong. The view that all human beings have the right to life might be given a non-religious defense of at least two kinds. On the one hand, it might be defended on the grounds that it is just moral common sense, but, if so, it is subject to Singer's speciesism objection. On the other hand, it can be defended by appeal to BLG. However, the Lee-George defense of the major premise of BLG seems unsatisfactory. What should we conclude?

We certainly do not have grounds for concluding that abortion choice is morally permissible. The future of value argument is also a candidate for rejecting the moral permissibility of abortion. In addition, we should not conclude that the basic natural capacity for rational agency is (contra Singer) morally unimportant. Considerations other than those offered by Lee and George may be sufficient for showing that the basic natural capacity for rational agency underlies the right to life. Such considerations exist. In the next section I shall try to make them plausible.

1.3 The Moral Significance of the Basic, Natural Capacity to Exhibit Rational Agency

According to the future of value argument, it is wrong to kill *you,* reader, because killing you would deprive you of a future of value. Your future of value now is that set of experiences you will value in the remainder of your life. Your future of value does not, of course, exist now. In addition if you are killed now, your future of value will not exist in the future, although, according to the future of value argument, you, in the moment before now, had a future of value. It could hardly be otherwise if killing you *deprives* you of your future of value. To deprive someone of something is to take it away from them. When something has been taken away from someone they don't have it.

Here is a little puzzle. What can talk of futures of value come to if such talk is not about what actually is the case? One might argue that, on the supposition that you are killed now, your future of value does not exist *now* (because it is *future*) and it will not exist in the future (because you have just been killed). True claims must be based on what is actually the case. But if a so-called future of value is about neither what is now the case nor what will be actually the case, then talk of futures of value is, to put the point more decorously than usual, mere hot air.

To this little puzzle there is a general response. According to the future of value account of the wrongness of killing, killing is one species of harm; indeed, it is one of the most severe kinds of harm. Harming someone involves making them worse off than they otherwise would have been. Unless there is some basis for making true statements about how one's life would have gone in the absence of the harm, the harm claim is just hot air. In short, we need to have some basis for making a comparison between the state of someone who has been harmed and the state she would have been in if she had not been harmed. The state she would have been in is neither the actual state she was in after she was harmed nor the state she was in before she was harmed. What can be the basis for such talk?

We know a good deal about how life would have gone for us if a particular harm had not occurred. If I catch a cold and do not feel well as a result, I judge that catching the cold has harmed me and I make that judgment because I judge how life would have gone for me in the absence of the cold. If I am fired from my job, I judge that I have been harmed and the basis for that judgment is a good general idea of how life would have gone for me if I had not been fired. Someone who dies prematurely has a worse life than someone who lives out a natural life because, in general, a shorter life is worse than a longer life. A shorter life is worse, in general, because it contains less of value. If we speak of this from the perspective of the time of a killing, then one's shorter life is worse because it does not contain one's future of value.

To say now that you now have a future of value is to talk now about your present potentiality. We know a good deal about how someone's life would have gone if he had not died prematurely. We know this because we know a good deal about normal human lives. We are biological organisms. Biological organisms have life spans and we know a good deal about the human life span from the study of biology, from the study of biography, from reading novels, from our knowledge of our parents, grandparents and friends. We know, in short, a great deal about the natural history of human beings. This knowledge is the basis for claims about the futures of value of human beings.

To know about the natural history of human beings, as opposed to the natural history of German shepherds, or the natural history of oak trees, or the natural history of mosquitoes, is to know about the kind of individuals humans are. It is to know about their essential natures. Just as the capacity of table salt to dissolve in water is based on its essential nature, the basic natural capacity of human beings to manifest signs of rational agency is a dispositional property of human beings and that dispositional property is based on the kind of things human beings are. Thus, once we understand how killing a human being is a basic harm, we are drawn into, after analysis, realizing that our understanding of this harm typically involves understanding what we essentially are, what our basic natural capacities are. All individuals who have the basic natural capacity for rational agency are individuals who have futures of value. Accordingly, the future of value view can provide us with a defense of the major premise of BLG. The major premise of BLG is true.

1.4 Some Flies in the Ointment

If one is primarily concerned with the ethics of abortion, the above analysis may seem entirely plausible. However, there are possible problems. Consider some particular fetus called "Oscar". How do we know that Oscar would have lived out his natural life span if he were not aborted? We don't know that, and so the above analysis must be qualified by considerations of the expected value of Oscar's life, or by some other line of analysis that takes into account these uncertainties. In my view, these modifications do not undermine the future of value view or the major premise of BLG. Discussion in this essay of just exactly what these modifications should be would tax the reader unreasonably.[8]

Other difficult issues bear more directly on this analysis. Suppose you are diagnosed with glioblastoma. Suppose your oncologist tells you that your median life expectancy is seventeen months. Now suppose that someone kills you tomorrow. Plainly your being killed did not deprive you of a normal human life span, for the glioblastoma robbed you of that. Therefore, what it is about a particular human being at a particular time that underwrites the truth of claims about that particular human being's future of value is not the natural life span of human beings in general, but the state of that particular human being just before she is killed. That state may not underwrite your capacity for a normal human life span. However, that state will underwrite your basic, natural capacity for rationality and self-awareness, and therefore underwrite your future of value. This issue poses no real difficulty for the major premise of BLG.

1.5 The Minor Premise of BLG

The preceding analysis will allow us to state BLG more explicitly. It is:

1. All individuals who possess the basic natural capacity for rational agency have the right to life.
2. All human beings are individuals who possess the basic natural capacity for rational agency.
3. Therefore, all human beings have the right to life.

The argument of the preceding sections has concerned the defense of the major premise of BLG, but not directly the defense of the substantial identity view. A defense of the substantial identity view also requires a defense of the minor premise of BLG. This premise is harder to defend than the major premise.

Suppose, for some reason, you fall into an irreversible vegetative state, a state in which you are, and always will be, incapable of either awareness or purposive movement. Why would it be wrong to end your life? According to the future of value account of the wrongness of killing, the standard reason why ending the life of a human being is wrong is that ending her life would deprive her of a future of value. If you are in an irreversible vegetative state, you lack a future of value. Therefore, unless there is some other reason why it would be wrong to end your life, ending your life would not be wrong. Would it have been wrong to end your life according to the substantial identity account of the wrongness of killing? According to the *conclusion* of BLG, according to the substantial identity view itself, it would have been wrong to kill you because you are a human being and it is always wrong intentionally to end the life of a human being. However, anyone in an irreversible vegetative state lacks the basic natural capacity for rational agency. Since the major premise of BLG is what ultimately underwrites the wrongness of killing you, it is most unclear why it is wrong to end your life, even if the basic principle on which Beckwith, Lee, and George base the wrongness of killing is true![9] The problem here is that the minor premise of BLG is false. Therefore, BLG fails to provide support for the substantial identity view. This analysis suggests that it is of the very last importance to distinguish between the major premise of BLG and the substantial identity view itself.

Other cases show the importance of this distinction. Suppose we ask whether or not anencephalics are victimized if their lives are ended intentionally. On the one hand, since anencephalics lack the basic, natural capacity for rational agency the major premise of BLG does not imply that killing anencephalics is wrong. Therefore, the basic principle on which the wrongness of killing is based, according to Beckwith, Lee, and George, does not imply that killing anencephalics is wrong. On the other hand, because anencephalics are members of our species, the conclusion of BLG does indeed underwrite the wrongness of ending the lives of anencephalics. This issue arises whether the anencephalics we consider are fetuses or infants. The problem here, put another way, is that anencephalics are counterexamples to the minor premise of BLG.

Donation of vital organs by brain dead human donors and the substantial identity account do not easily coexist. Consider human beings on ventilator support who have suffered total and irreversible loss of brain function, but whose other organs are viable. According to the orthodox definition of death all such individuals are dead. According to the moral common sense that supports the substantial identity view, because, and only because, such "merely brain dead" individuals are dead, their vital organs can be transplanted, with their consent (or with the consent of next of kin), into those who are in need of an organ transplant. However, there now is, in the bioethics literature, compelling criticism of the orthodox definition of death.[10] Its essential nature is quite simple.

Biological organisms are, by definition, living. For an organism to be living is for the organism as a whole to exhibit integrated biological function. An organism is dead if and only if it is no longer living. Merely brain dead human biological organisms are also human biological organisms that exhibit integrated biological function, although they require life support.[11] Indeed, if they did not exhibit such integrated biological function, then warm ischemia in the organs in such organisms would result in organ necrosis and unsuitability for transplant. Is it so clear that a merely totally brain dead human individual can exhibit integrated biological function? Such individuals exhibit integrated respiratory function, exhibit integrated circulatory function, can deliver live offspring if pregnant, exhibit integrated excretory function, digest and deliver nutrients throughout the body and exhibit many other functions that are functions of the organism as a whole. That these functions continue only with the artificial support of a ventilator does not show that such organisms are not alive, for human beings, such as juvenile diabetics and most fetuses(!), continue to live only with the aid of life support.

Conclusive arguments exist for the view that the merely brain dead are, contrary to contemporary orthodox legal doctrine, living human beings. What are the implications of this? On the one hand, the substantial identity view apparently implies that removing the vital organs of a merely brain dead human being is wrongful killing. This, of course, is incompatible with the moral permissibility of most vital organ donation. On the other hand, the major premise of BLG is compatible with the moral permissibility of removing the vital organs of a merely brain dead human being, for a merely brain dead human being lacks the basic natural capacity for rational agency. This discrepancy points again to a difficulty with the minor premise of BLG and the very real tension in the BLG defense of the substantial identity view.

Lee and George have tried to deal with the substantial identity view's apparent prohibition of vital organ donation in terms of a very brief suggestion that vital organ donation could be permissible using a variant of the donation after cardiac death protocol (Lee and George, 2008, 169). Whether this suggestion is either ethically or medically defensible requires far more discussion than Lee and George have offered or than is compatible with keeping this essay to a reasonable length. At any rate it is safe to conclude that the minor premise of BLG is false, and that, as a consequence, a defense of the substantial identity view in terms of BLG fails. It is worth noting that the truth of BLG's basic moral principle is entirely compatible with the usual sort of vital organ donation.

1.6 A Possible Defense of the Substantial Identity View

According to the future of value account, one's future of value is constituted by the *experiences* one has the potential to value in the remainder of one's life. This characterization of the future of value account could be altered. One might say that the potential *goods* in the remainder of one's life constitute one's future of value. The difference between the two versions of a future of value is that the latter version leaves room for goods in one's life that lack experiential content. Let us call the former version of the future of value view "the experiential version" and the latter version of the future of value view "the goods version". Let us now suppose that the goods version of the future of value view is correct. Let us suppose further that human biological life by itself is an intrinsic good. These two claims provide a basis for the inference that individuals in an irreversible vegetative state and that anencephalics have futures of value. If this analytical strategy can be defended, then the future of value account of the wrongness of killing seems to support the substantial identity account.[12]

If human biological life by itself is an intrinsic good, then how should the relationship between the goods version of the future of value account and the minor premise of BLG be characterized? Anyone with a basic natural capacity for rational agency has a future of value. However, some individuals, such as anencephalics, who have a future of value (goods version) lack the basic natural capacity for rational agency.

What does it mean to say that human biological life is an intrinsic good? Everyone reading this essay has a life that has intrinsic value, that is, your life has value in and of itself. This does not mean that one's life does not have value for others, but only that its value for others does not exhaust its value. The standard view is that the intrinsic value of a human life is, by itself, sufficient to make it presumptively seriously wrong to end another's life.

When we talk about our lives, what exactly are we talking about? "Life" is ambiguous. When I think of what has gone on in my life, I think of my experiences, activities, memories, projects, and such things. These things all have mental constituents. A life of this sort can be called a "biographical life".

However, one can think of what has gone on in one's life in an entirely different way. Imagine a physiologist or a physician writing about you and recording the events in your biological life. She might write about your heart rate, your blood pressure, your kidney function, your liver function, your muscle mass and many other such things at various times in various days of your life. She would be writing about your biological life.[13]

Does one's biological life have intrinsic value? My being biologically alive is a necessary condition of my having the experiences I value in my biographical life. Furthermore, the experiences I can have in my biographical life may be limited if my biological function is poor. Therefore, my biological life has great instrumental value. Indeed, it is reasonable to say that biological life is the sort of thing Rawls called "a primary good". Like money, it is useful for obtaining many of the things in life that I value for their own sake. However, this general usefulness does not make biological life intrinsically good any more than it makes money intrinsically good.

Of course, it does not follow from this that human biological life lacks intrinsic value, any more than it follows from the fact that we are useful to others, and therefore have instrumental value for others, that our lives, or that we, lack intrinsic value. However, the preceding considerations suggest that an argument for the claim that mere human biological life has intrinsic value is needed. Plainly human biological life has great value, since its instrumental value is both so evident and so important. However, this fact does not support the substantial identity account. Someone in irreversible vegetative state does not have a life that makes possible the experiences of life that all of us value for their own sakes. The biological lives of such people, unlike ours, do not have instrumental value for them. Therefore, if the biological lives of such people have value for them, the value they have must be intrinsic value. If this is not the case, then the substantial identity view is false.

Does the life of someone in irreversible vegetative state have value for the person whose life it is? One way of approaching an answer to this question is by doing empirical research. We could ask people whether they believe that, if they were irreversibly unconscious, they would lose something of value to them if their lives were ended. I doubt that people would value such a life, but I have not done good empirical research into this question. I conjecture that the deep reason why many people thought it was not wrong intentionally to end the life of Terri Schiavo is that they believed that ending her life would not victimize her by depriving her of something of value to her. For myself, I cannot imagine why I would care if my life were ended if I were to fall into an irreversible vegetative state.

Another way of answering this question is by looking at the relevant arguments. Lee and George have offered an argument for the view that life in an irreversible vegetative state is intrinsically good. Here it is:

> ...what makes a condition or activity intrinsically valuable, worth pursuing for its own sake, is that it is *fulfilling*. But it makes no sense to hold that the fulfillment of an entity is intrinsically valuable, and yet the entity itself is not. The entity itself cannot be viewed as a mere instrumental good or as a mere condition for the fulfillment or perfection of that entity. Thus my genuine good includes my *being* as well as my *full*-being (Lee and George, 2008, 161).

Lee and George are quite correct. A state of me that is good for me, that is a fulfillment of me, must, of course, include me. My doing philosophy, my playing the piano, my having a nice dinner with friends all include me. It does not follow that if I am not (indeed, cannot be) fulfilled by experiences that are good for me that I have any value at all. The conclusion that Lee and George actually establish is plainly true, but it is not the conclusion that a human *without* the potential for any fulfilling experience is intrinsically valuable.

Lee and George also defend their view by saying, "But what I am is a living bodily entity and the thing which I am is intrinsically valuable, so it follows that this bodily entity itself is intrinsically valuable (Lee and George, 2008, 161)" No one (except the few dualists among us) doubts that Lee and George are living bodily entities. No one who thinks they are living bodily entities doubts that they are intrinsically valuable, whether they are intrinsically valuable in virtue of their essential properties or their accidental properties. Does it follow that those two bodily entities themselves are intrinsically valuable *independently* of any properties they may possess? One needs an argument here that accidental properties are not sufficient to give one value. *This* argument does not even attempt to establish *that.* Therefore, Lee and George have not shown that a human biological life, in the absence of the capacity for a biographical life, has intrinsic value. Accordingly, they have not rescued the substantial identity account.

Interestingly, however, sound arguments exist for the view that human biological life is an intrinsic good. The biological function of a human organism in an irreversible vegetative state can be better or worse. Such an organism can function well or poorly. It can be harmed or benefited. Such an organism has, indeed, a good of its own. This truth will not provide aid and comfort to substantial identity theorists because *all* biological organisms have a good of their own. Biological organisms have a teleological nature and can flourish or wither. This is a fact about mosquitoes and oak trees as well as about human biological organisms. This fact will not underwrite a duty not to terminate any biological life (unless it underwrites a duty not to swat mosquitoes) and will not, therefore, support the substantial identity view. Therefore, a goods version of the future of value view will not support the substantial identity account of the wrongness of killing.

1.7 Summary and Conclusion

Interesting issues concerning the substantial identity account of the wrongness of killing and the future of value account have been neglected in this essay. First, I have not, as I pointed out earlier, fully explored strategies for reconciling the substantial identity account with a robust vital organ transplantation program. Second, I have not explored whether the substantial identity view, as its proponents claim, supports or is fully compatible with a robust doctrine of human equality. There are reasons for doubt on this topic. The difficult issue is whether, in cases of the refusal of extraordinary means of preserving life in health care contexts, the burdens and benefits analysis on the basis of which witholding an intervention is judged morally permissible always can (or does) proceed independently of the diminished value of the life of a dying, or gravely ill, patient. If it does not, or cannot, then one must wonder about how this is compatible with the doctrine of the equal worth of all human life that is supposed to be a feature of the substantial identity view. Third, I have not explored whether or not the major premise of BLG and/or the version of the future of value view to which it is (almost) fully equivalent is compatible with the full respect for the lives of living human beings with severely diminished mental lives, either as a result of life-long retardation or as a result of end of life dementia. Perhaps such lives can have value to their possessors and yet lack the capacity for rational agency. There is plainly a difficulty here and an exploration of it is not without interest. A virtue, of course, of the substantial identity view (but not the major premise of BLG!) is that such a difficulty does not arise. Both the major premise of BLG and the future of value view are subject to this difficulty, if it is a real difficulty. Hence, if this difficulty is a real difficulty, then substantial identity proponents will have to give up BLG. The problem then for them concerns how they would *defend* the substantial identity view. Fourth, I have not explored the implications of the major premise of BLG for those difficult situations in which intentionally ending the life of a human being seems to be not only morally permissible, but morally required. I invite you to consider the following case. You are a soldier in the trenches in World War I. A fellow soldier has suffered horrendous wounds from which he will surely die within hours. He is howling for morphine to control his pain. There is no morphine or other medical aid close by. Why wouldn't you shoot him? Notice that here is a case in which the capacity for rational agency has not been lost, but in which the soldier lacks a future of value. Notice also that here is a case in which the future of value view and the major premise of BLG do not yield the same results. Fifth, I have not carefully explored some major deep issues about the wrongness of killing that go to the heart of the project of philosophical ethics. The future of value view starts from an account of a harm that we all take to be a morally significant harm and moves to a view concerning moral status. The substantial identity view seems to start with a view of what has great intrinsic value, of what has moral status. I believe that the former approach is superior to the latter approach, but I certainly have not shown that. It also may be the case that these approaches, when explored carefully in a comparative way, are not as different as they may sometimes seem. Finally, I have not considered issues that arise from the fact that both the major premise and the conclusion of BLG are a good deal more specific than the future of value view. For all one finds in the various presentations of the future of value view, rabbits may have futures of value, at least those that have the potential to enjoy hopping and lettuce. Neither the major premise nor the conclusion of BLG makes ending the life of a rabbit wrong.

In spite of numerous loose ends surrounding this discussion, some fairly clear themes have emerged. First, one might defend the substantial identity account of the wrongness of killing by appealing to the moral common sense of most people. This strategy won't do. It is open to the charge of speciesism. Second, one might defend the substantial identity view by appealing to religion. Such a defense is obviously insufficient. There are too many different religions. Third, one might defend the substantial identity view by claiming that alternative views of the wrongness of killing are arbitrary in some way. This is insufficient. If we are going to defend a view that is incompatible with reproductive freedom, then we need a *positive* account of the wrongness of killing that *explains* why killing a standard human being is wrong. Fourth, one might defend the substantial identity view by appealing to BLG. BLG is not open to charge of speciesism. A defense of BLG must involve a defense of a basic, natural capacity interpretation of its major premise. The Lee-George defense of a basic natural capacity interpretation of BLG's major premise is subject to difficulties. However, an adequate defense of BLG's major premise that appeals to the future of value account of the wrongness of killing is available. BLG's minor premise is its weak link. It is subject to counterexamples. Therefore, the BLG defense of the substantial identity view is unsound. Fifth, one might attempt to defend the substantial identity view by appeal to the goods version of the future of value view. This defense works only if it is true that mere biological human life is intrinsically valuable. The Lee-George defense of that view is subject to difficulties. There are arguments that mere human biological life is an intrinsic good, but these arguments are not sufficient to show that the intrinsic good of human biological life has any moral significance. Therefore, this good cannot be the basis for the right to life of a human being. It seems, therefore, that all of the defenses of the substantial identity view are subject to serious difficulties.

Of course, none of the above conclusions lend aid and comfort to those who are in favor of abortion choice. The view that ending the life of a fetus is (almost always) a wrongful killing can be justified in terms of the future of value account of the wrongness of killing or in terms of the major premise of BLG.

Notes

1. Some of these counterexamples can be found in Tooley, 1972.
2. In my view this is a mistake.
3. Jeff McMahan (2002) is the most notable example.
4. I shall neglect entirely arguments concerning infanticide offered by Lee and George (2008, 136) and Beckwith (2007, 139) because they divert us from the issue of *why* it is wrong to kill human beings, because their analysis would make a long essay even longer, because they appeal to cases where intuitions differ, and for other reasons.
5. Stretton (2004) apparently has made this argument, although because he illustrates it with a claim about chimpanzees, I'm not certain.
6. It is worth noting that this threshold point undermines the Lee-George continuum argument as well.
7. Beckwith (2007, 134) makes this claim.
8. A wonderful discussion of some of these difficult issues can be found in McMahan, 2002, Chapter 2.
9. I neglect the possibility that you left money in your will for your care in a reversible vegetative state or that your parents want to care for you at home.
10. In my view the most thorough, and most devastating, critique of the orthodox definition of death has been set out in Shewmon, 2001.
11. This is not an arbitrary assumption in the context of this essay. Think of the criteria for saying that embryos and fetuses are living human beings.
12. Boyle (1989, 2008) has defended a version of this strategy in a number of places.
13. The distinction between biological life and biographical life is due to Rachels, 1986, 24–25.

References

Beckwith, F.P. 2007. *Defending life: A moral and legal case against abortion choice*. Cambridge, MA: Cambridge University Press.

Boyle, J.M. 1979. That the fetus should be considered a legal person. *The American Journal of Jurisprudence* 59:59–71.

Boyle, J.M. 1989. Sanctity of life and suicide: Tensions and developments within common morality. In *Suicide and euthanasia*, ed. B. Brody, 221–250. Dordrecht, The Netherlands: Kluwer.

Boyle, J.M. 2008. Towards ethical guidelines for the use of artificial nutrition and hydration. In *Artificial nutrition and hydration: The new Catholic debate*, ed. C. Tollefsen, 111–121. Dordrecht, The Netherlands: Springer.

Brown, M. 2000. The morality of abortion and the deprivation of futures. *The Journal of Medical Ethics* 26:103–107.

DeGrazia, D. 2005. *Human identity and bioethics*. New York, NY: Cambridge University Press.

Dworkin, R. 1993. *Life's Dominion: An argument about abortion, euthanasia, and individual freedom*. New York, NY: Alfred A. Knopf.

Harris, J. 1999. The concept of the person and the value of life. *Journal of the Kennedy Institute of Ethics* 9:293–308.

Lee, P. 2004. The pro-life argument from substantial identity: A defense. *Bioethics* 18:249–263.
Lee, P., and R.P. George. 2008. *Body-self dualism in contemporary ethics and politics*. Cambridge, MA: Cambridge University Press.
Marquis, D. 1989. Why abortion is immoral. *The Journal of Philosophy* 86:138–202.
McMahan, J. 2002. *The ethics of killing: Problems at the margins of life*. Oxford: Oxford University Press.
Rachels, J. 1986. *The end of life*. Oxford: Oxford University Press.
Reiman, J. 1999. *Abortion and the way we value human life*. Lanham, MD: Rowman and Littlefield.
Shewmon, D.A. 2001. The brain and somatic integration: Insights into the standard biological rationale for equating 'Brain Death' with death. *Journal of Medicine and Philosophy* 26: 457–478.
Singer, P. 1979. *Practical ethics*. Cambridge, MA: Cambridge University Press.
Steinbock, B. 1992. *Life before birth*. New York, NY: Oxford University Press.
Stretton, D. 2004. Essential properties and the right to life: A response to lee. *Bioethics* 18: 254–281.
Tooley, M. 1972. Abortion and infanticide. *Philosophy and Public Affairs* 2:37–65.

Chapter 2
Substantial Identity, Rational Nature, and the Right to Life

Patrick Lee

2.1 The Substantial Identity View

In various places many natural law theorists, including Joseph Boyle, have defended the position that every human being is valuable as a subject of rights in virtue of the fundamental or substantial kind of being he or she is, rather than in virtue of additional, accidental attributes, that may be acquired (and then lost) later in life.[1] From this position it follows that a human being is valuable as a subject of rights from the moment he or she comes to be, and remains valuable as a subject of rights as long as he or she continues to live. The morally relevant substantial kind that qualifies one as a subject of rights—we (natural law theorists) have argued—is a being with a rational nature (Lee and George, 2008b). Thus, if there are other rational animals besides humans, then they also are subjects of rights.[2]

A brief rationale for this position is as follows. In moral deliberation we directly apprehend that our genuine fulfillment (various basic goods) is worth pursuing and protecting. The fulfillment grasped as worthy of pursuit is not just one's own fulfillment as an individual, but the fulfillment of oneself and of others with whom one can have personal communion, those with whom one can rationally cooperate as co-agents. Moreover, the fulfillment of these subjects is worthy of pursuit and protection during those times that they are not conscious, and so it is not just *actual* co-agents whose fulfillment is worth pursuing but also those entities that have the basic capacity (even if not immediately exercisable) of being rational agents. Subjects with a capacity for rational pursuit of fulfillment and cooperation are worthy of full moral respect during all times that they exist.

This position is in sharp contrast to the view that in order to be valuable as a subject of rights an entity must have, not just a rational nature, but an immediately exercisable capacity (or nearly immediately exercisable capacity) for rationality, self-consciousness, or the like.[3] The standard rationale for this position is that rights are based on interests or self-conscious desires, and thus in order to possess

P. Lee (✉)
Franciscan University of Steubenville, Steubenville, OH, USA
e-mail: plee512@gmail.com

C. Tollefsen (ed.), *Bioethics with Liberty and Justice*, Philosophy and Medicine 110,
DOI 10.1007/978-90-481-9791-0_2,

rights one must either now have, or have had in the past, immediately exercisable capacities for self-conscious desires.

A third approach to the basis of being a subject of rights is that of Don Marquis. Marquis argues that the basis for personal dignity is the possession of a future of value, a future that includes a positive set of mental experiences. This position is much closer to the substantial identity view than to the immediately exercisable capacities view: on Marquis's position almost all human beings from conception on have basic rights. Still, Marquis' position implies that some human beings do not have rights, since not all human beings have, according to him, a future of value (Marquis, 1989). In his contribution to this volume, Marquis has criticized the substantial identity view as this has been articulated by Boyle, myself, and our colleagues, contrasting our view with that of Peter Singer and with his (Marquis's) own.[4] Marquis's paper greatly clarifies the issue and advances the debate, pressing for further precision. In this paper I will try to clarify the substantial identity position and reply to Marquis's criticisms of it.

2.2 Contrast with Singer's View

Marquis summarizes the basic argument for the substantial identity view (advanced by such philosophers as Robert P. George, Francis Beckwith, and myself) in the following syllogism, which he calls the "BLG" argument (after Beckwith, Lee, and George):

1. All individuals who are rational beings are individuals who have the right to life.
2. All human beings are individuals who are rational beings.
3. Therefore, all human beings are individuals who have the right to life (Marquis, 2011).

He points out that Peter Singer (and others) agrees with the first premise but disagrees with 2. Singer holds that only beings that have an *immediately exercisable capacity* for rationality are rational beings, and thus are subsumed under the first premise. By contrast, proponents of the substantial identity view hold that every entity that has a *basic natural capacity* for rationality qualifies as a rational being, and thus has a right to life. Moreover, every human being is a being with a rational nature, and thus has a basic right to life.[5] A capacity is an ability to perform an action. If one has the constitution enabling one to perform that action at the present time, then one has an immediately exercisable capacity, or a first-order capacity (the capacity to do X). If one has the constitution now that will enable one to develop the constitution that enables one to perform the action later, then one has a basic, natural capacity, or a second-order capacity (a capacity to develop a capacity to do X). The substantial identity view differs from that of Singer (and Dworkin, Steinbock, and others) as holding that a basic natural capacity for rationality (more precisely: being a rational agent) is sufficient for being a subject of rights.[6]

One difficulty with Singer's view is that—as Marquis points out—it seems that an individual can be harmed (deprived of a good) before he or she is conscious. Since you and I are human beings, and a human being is essentially a physical organism, then you and I were alive as embryos and fetuses. Thus, you and I could at a time prior to our being conscious (and thus prior to having an immediately exercisable capacity for rationality), have been deprived of a valuable future. (On this the substantial identity view and the future of value view agree.)[7]

A second difficulty with Singer's view is that his rationale for it is what Marquis calls a positive attitude account. On this account, the value of a future state of an individual derives from that individual's positive attitude toward that future state. S's being alive at a future time is valuable and to be respected (promoted, protected, etc) on the grounds that S now desires, cares about, takes an interest in, or has desired, cared about, taken an interest in, that future. But on this view people who are so clinically depressed or brainwashed as not to desire their future life, would not have a right to life, which is false.

Moreover, the positive attitude account puts the cart before the horse. Conditions are not worthwhile because they are desired; rather, they are desirable because they are worthwhile, because they have whatever it takes to make something worth being desired. Clearly, some of our desires are bad and some are merely whimsical. Since worthless objects are sometimes desired, it follows that being desired cannot be what makes an object to be worthwhile. So, prior to being desired, the object of desire must have something in it which makes it *fitting* or *suitable* to being desired. What makes a thing good does not consist in its being the satisfaction of desires or preferences; rather, desires and preferences are rational only if they are in line with what is genuinely good. So, a state of affairs should be desired or cared for if it is inherently valuable. A condition's being valuable makes desires for it reasonable, not vice versa (Lee and George, 2008a, Chapter 3).

A third difficulty with Singer's view is that the proposed criterion for possession of basic rights is actually a certain degree of a property along a continuum of development of that property. Singer's position (and others like it) involves an arbitrary selection of a certain degree of a property as the basis for basic rights, while ignoring the radical (and substantial) difference that occurs with the coming to be of an individual human organism (an individual with a developmental trajectory toward the mature stage of a human, where this same organism will then reason, make free choices, and so on). Marquis calls this difficulty against Singer's position "the continuum argument," but rejects it. If this argument is sound, then it also tells against Marquis's own position (since the argument also tells against the future of value criterion as Marquis interprets it). In the next section I will try to clarify this argument and answer Marquis's objections it, and in subsequent sections examine further objections from Marquis.

2.3 Properties, Continuity, and Arbitrariness in Selecting the Basis for Being a Subject of Rights

Several arguments have been advanced to support the substantial identity position. A central one is what Marquis calls the "continuity argument." But it would better be called a discontinuity-and-continuity argument. We treat beings with basic rights in a radically different way than we treat beings we think do not have rights. Most people believe it is permissible to use, experiment on, dismember, and even kill, for our own purposes, beings that do not have rights, even though we think we should not cause unnecessary pain to them when doing so. By contrast, we believe, that, at least in general, we should not intentionally kill beings with rights, and even that we should treat them as we would have them treat us.[8] But it would seem that such a radical difference in the way *we treat* different classes of beings should be based on some radical difference *in reality* between those classes of beings. Now, there *is* a radical difference between substances with a rational nature, on the one hand, and other substances.[9] In particular, considering the beginning of life, we see there is a radical difference in kind between a sperm and an ovum on the one hand, and an embryo, from the zygote stage onward, on the other. The former are sex cells, parts of larger organisms (the paternal and maternal organisms respectively); the latter is a human being at his or her early stage of development, and is the same substantial entity that will later be born, crawl, walk, talk, and eventually reason and shape his or her own life. But once the human being comes to be, there is only a difference in degree between him or her and the subsequent stages of that same human being. There is no radical difference between a human being at one stage of his or her development, and that same human being at another stage of development.[10] Hence those who propose an accidental attribute, such as an immediately exercisable capacity for self-consciousness, as the basis for being a subject of rights, ignore a morally relevant radical difference in kind between two classes of beings, and have instead selected a mere difference in degree in the development of a human being for the criterion of basic rights. But the selection of *this* degree of development rather than *that* degree, is arbitrary, and therefore, unjust.

It is true that in some areas we must make an arbitrary selection of when to treat different people differently—for example, in determining when to count someone as intoxicated, or at what age to allow someone to drive an automobile (as Marquis points out). But in those cases there is no morally relevant radical difference one could select. Here, on the contrary, regarding basic rights (as opposed to, say, specific rights to perform specific actions) there *is* a nearby and morally relevant radical difference. But proponents of the position that being a person is an accidental characteristic ignore that difference and pick out a mere quantitative difference as the basis for radically different types of treatment. Hence it is more reasonable to base the right to life on the *substantial nature* of the entity, rather than on developed capacities, such as an immediately exercisable capacity for self-consciousness, or self-conscious desires.

Marquis rejects this argument. He interprets it as saying the following: there are two kinds of capacities: basic natural capacities, and immediately exercisable capacities; to base basic rights on the latter is to base them on something that varies in degree and leads to arbitrary line-drawing. Marquis then asks: "What precisely is this kind of capacity of which one has more as one develops and, conversely, of which one has less the shorter the time that has elapsed since one's conception" (Marquis, 2011, 9)? And his answer is, first, that it cannot be the basic, natural capacity for rationality, since that capacity does not vary in degree over time.[11] And, second, it cannot be the immediately exercisable capacity, since that capacity is not present at all in fetuses and embryos. So, Marquis concludes: "The feature that

Lee and George presumably are talking about [and reject as a basis for possession of basic rights] that develops so that there is more of it as the gestational age of a fetus increases does not exist." In short, according to Marquis "the continuity argument" is incoherent—there is no capacity, whether basic, natural or immediately exercisable, possessed by all human beings throughout all of their lives, that varies in continuous degrees. So, this argument (says Marquis) fails to show that Singer et al.'s choice of one feature rather than another involves an arbitrary line-drawing.

But the argument here needs to be clarified.[12] Specifically: speaking of two kinds of capacities, while convenient, is misleading. What is referred to as an "immediately exercisable capacity" for rationality is not actually a new and different capacity from the corresponding basic natural capacity. Rather, it is only a *degree of actualization or development* of the basic capacity for rationality that exists in the human being from the beginning. A capacity such as that for consciousness is a power to perform a specific type of action. It is grounded in the constitution of the organism. The capacity develops and comes closer to being the performance of that action, with the development of the constitution of the organism; but in a living being, the transition from the basic natural capacity to perform an action characteristic of living beings, on the one hand, to performance of that action, on the other, is just the development of the basic power that the organism possessed from its beginning. Thus, the feature Singer and others have actually based the right to life on, is *the development of* the one potentiality for rationality (and deliberate choice) that exists in the organism all along. So, the problem with basing the right to life on an accidental feature that varies in degrees remains: basing the possession of basic rights on the immediately exercisable capacity for rationality is just the selection—and necessarily an arbitrary selection—of a certain degree of development of the one capacity for rationality that the human being possesses from his or her beginning.[13]

One might object that when a human organism reaches the stage at which he or she has the immediately exercisable capacity for consciousness, there in fact does occur a radical change, even though it is produced by gradual development. By analogy, a gradual change in the temperature of water gives rise to a qualitative change from liquid to vapor.[14] However, while there *is* a qualitative change in the development of a human organism with respect to consciousness, that change does not occur until the attainment of *actual* consciousness. And no one will claim that actual consciousness is the criterion for having basic rights—otherwise human beings who are asleep or temporarily unconscious would lack basic rights.[15] In the analogy of water, the transition from liquid to gas is an *effect* of other gradual changes. By contrast, in the developing human, the immediately exercisable capacity for consciousness is not a distinct effect of the development of the basic natural capacity for consciousness; it is just a further degree of development of that same capacity. While there is a qualitative difference between the capacity for consciousness and actually being conscious (between having a capacity for a behavior property and actually doing it), there is no qualitative difference between having a capacity for consciousness and having a more developed capacity for consciousness.[16]

Marquis claims that the fact that the immediately exercisable capacity to perform characteristically human mental functions develops by degrees *after* we are born poses no difficulty for Singer's view. As Marquis recounts Singer's view, once the organism has that capacity, even though the capacity continues to develop afterwards, the human organism is a person. And according to Marquis, Singer's way of "drawing the line between individuals who are persons and individuals who are not does not involve drawing a line at some arbitrary time during the development of the immediately exercisable capacity for rationality" (Marquis, 2011, 9).

However, since there is no distinct immediately exercisable capacity whose beginning could be a non-arbitrary criterion for being a rights-bearer, Singer's position *does* involve an arbitrary line-drawing. For it is arbitrary where in the development of the one real capacity for rationality one will draw the line between non-persons and persons. By contrast, for the substantial identity view there is a non-arbitrary line: the beginning to exist of the substance with a rational nature: before that what exists are only cells (sperm and ovum) that could join to produce a new substantial being (a new human being); after that what exists is a self-developing, whole, though obviously immature, human being.

2.4 Equality and Threshold Properties

Marquis objects also to another argument we have advanced to support the substantial identity view, which he calls "the equality argument." We have argued that if one bases being a subject of basic rights on an accidental characteristic (or characteristics), and that characteristic varies in degrees, it seems to follow that possession of basic rights will also vary in degrees. Some subjects of rights will be "more equal" than others. Marquis first objects that the basic natural capacity for rationality also varies in degrees, and so it is not superior to accidental characteristics as a criterion for basic rights—specifically, having a future of value—in this respect (Marquis, 2011, 10). In other words, we can hardly argue against other views (including his future of value view) on the grounds that the other proposed criteria vary in degrees, when the criterion we ourselves propose also varies in degrees.

However, there is a difference between holding that possession of *a property or nature itself* qualifies one as possessing basic rights, and holding that possession *of a certain degree* or *development of* that property qualifies one as having basic rights. We hold the former (regarding a substantial nature), and on that basis it makes sense to hold that having a basic natural capacity for rationality and deliberate choice, that is, a substantial rational nature, means that one has full more worth, that is, is a subject of rights. But to hold that possession of a property to an nth degree, rather than possession of the property itself, is what qualifies one as possessing basic rights, is indeed subject to a slippery slope (as well as being arbitrary). For why should the nth degree qualify one as having rights? Why not the nth +1 degree, or the nth +2 degrees, etc?

An individual's having a rational nature means he or she is inherently—that is, in virtue of what he or she is, not in virtue of accidental characteristics—oriented to conceptual thought and free choice, thus, to shaping his or her own life, constituting how good or bad a person he or she will be by his or her free choices. Having a nature orienting one to shaping his or her own life has no degrees; either a being has that or not. If a being has that then one morally ought to respect that orientation (and that will mean not intentionally killing him or her, not enslaving him or her, and even, treating this individual as one would have him or her treat oneself).The criterion is not: ability to reason, or ability to be self-aware, which do vary in degrees. Rather, the criterion is: being an individual of the substantial kind characterized by having the natural capacity to reason and make free choices. In this there are no degrees: one either is or is not an individual with a rational nature.

In one fundamental respect all rational beings, all persons, are equal, namely, each is oriented to shaping his or her life morally well or morally badly. Other factors in an individual, that may facilitate his shaping his life well or make it more difficult to shape his life well, don't make a difference with respect to whether he has fundamental dignity and fundamental rights. Consider a human person who will have an IQ of 160 when mature vs. one with a much lower IQ—the high IQ person is not on that basis a better person. Moreover, even those who later *are* better persons, and therefore later do deserve better treatment in some contexts, do not on that account have more *fundamental* personal dignity than the morally less good or morally bad (who also could change their ways by free choice).

It also is true that not all persons have the same nature, even though they are the same in being persons. Non-human rational animals, if there are any (whether extra-terrestrial or not), angels, and God, are (in the sense we are using the term) rational beings, persons. These different natures might be unequal in various ways (angels are smarter than we are, and God is infinitely so), but those important differences are not connected to whether one has basic rights. Such basic rights are based on a respect in which all persons are equal (including humans, angels, and even God—God will not just use us for his own purposes). There is a radical difference between non-rational beings and all rational beings (humans, other truly rational animals if there are any, angels, God), and so that difference constitutes a non-arbitrary line. Above that, although they may be unequal in various ways, their inequalities are irrelevant to the possession of equal fundamental rights

Marquis considers something like what I have just said and objects to it. Robert George and I had said that the criterion for having basic rights is "having a nature that entails the capacity... for conceptual thought and free choice—and not *the development* of that natural basic capacity to some degree or other" (Lee and George, 2008a, 93). But, Marquis replies: "This misses the point. The objection to their equality argument is not that *the development* of the basic natural capacity can be a matter of degree (which it plainly is), but that *the basic natural capacity itself, whether developed to any particular degree or not* comes in degrees" (Marquis, 2011, 11). After all, he adds: "some people are naturally dim-witted and some are naturally very smart. Therefore, some people have a greater natural capacity for rational agency than others" (Marquis, 2011, 10).

But Marquis has still failed to distinguish between the *individual of a substantial kind* (in this case, the substantial kind characterized by its orientation toward rationality and deliberate choices), on the one hand, and other factors in an individual that may facilitate or impede the development of an individual's basic natural capacity for a rational mode of life, on the other hand.[17] These other factors are accidental in the sense that an individual might acquire such factors and not cease to be (or he may have come to be with such factors and still have been the same individual, and of the same kind, as without those factors): in other words, such factors are inessential. Thus, various genetic or other somatic factors may vary within me enabling me to develop more or less the basic capacities that belong to me in virtue of being the fundamental or substantial kind of being that I am. The point of the equality argument is that having basic rights is not tied to *these* factors, which vary in degree—which would involve a denial of *fundamental* equal dignity—but on having a substantial rational nature.

Having basic rights—moral immunity from being killed or enslaved, the right to be treated as others would have themselves be treated—follows upon being an individual with a rational nature, not on the possession of a degree of some property (such as the immediately exercisable capacity for reasoning), nor on factors in one's constitution that will aid or impede the more complete development of one's rational nature.

Other rights are based on the varying degree of ability to reason (scholarships, for example) or moral goodness (rewards and punishments), but having basic rights to begin with is based on being a certain type of substance or subject. Rights that are based on varying properties may themselves vary in degree. So, if there are equal basic rights—as opposed to rights that are merely conveniently treated as if equal—then what they are based on must also be equal.

Against the equality argument, Marquis also objects that the accidental properties view need not involve a denial of equal dignity, for the qualifying property can be viewed as a *threshold* property:

> The Lee-George equality argument fails for another reason as well. The alleged equality difficulty with Singer's view—or with their own—could be avoided by understanding the capacity for rationality (whichever kind of capacity one has in mind) as a threshold criterion for the right to life. . . .Therefore, it is not inconsistent to consider the capacity for rationality the criterion for the right to life, and to realize that rationality comes in degrees, and nevertheless, to hold the view that everyone who meets a certain rationality threshold has the right to life and has it equally (whether one adopts the Lee-George view or the Singer view) (Marquis, 2011, 11).

But of course one *can* select a certain degree of development of a capacity as a criterion for the right to life, and make that in effect a threshold property (as we do with legal intoxication, or driving age, for example). But that just is *deciding* to take the position that those who possess a given property to a certain degree will be treated as if they were equal in regard to fundamental rights. Our point in the equality argument is that the selection of a certain degree of a property as a criterion for being a subject of rights provides no reasonable basis for the idea that all individuals of a certain sort have equal fundamental rights. If having a property to an nth degree provides one with immunity against being killed, why not hold that one should have the nth +1 degree to provide one with that right plus immunity against being enslaved, and the nth +2 degree provides one with those rights plus the right to own property—why not have a sliding scale of basic rights? By contrast, the beginning to exist of a substance with a rational nature is a non-arbitrary basis for the possession of equal fundamental rights, since *from that very moment* on there exists an individual who has a nature orienting him or her to shape his or her own life. Every person, that is, every individual substance with a rational nature—whether that person be mature, powerful and fully alert, or on the contrary, immature, weak, dependent, and unconscious—is equally oriented to freely shaping a morally good or morally bad self.[18]

2.5 Biological Life and the Criterion for Basic Rights

The conclusion of the continuum (or discontinuity-continuity) argument and the equality argument is that human beings are valuable as subjects of rights in virtue of the kind of being they are, not in virtue of some additional characteristics. If *having a future like ours* is interpreted as an accidental property, then the continuum and equality arguments count against that criterion also. Still, Marquis might object as follows. He might deny that "mere biological life" is a real benefit. Suppose I am alive but I lack what Marquis would call a future of value. Suppose that I will be permanently unconscious. In that case, on the basis of the previous considerations one would say that I am still a bearer of rights and should not be intentionally deprived of basic goods. But Marquis might reply that biological life by itself is not a real benefit, and so even though I might in that case be intrinsically valuable, that would make no difference, since depriving me of biological life would not (in that case) violate my rights. Or he might argue that since I do not possess any future benefits (no future of value) then to say I am intrinsically valuable is empty, or false.

However, despite the initial appeal of this objection, I do not think it holds up to scrutiny. How does one determine which entities are subjects of rights, and which are not? Marquis contends that the basis for having rights is having a future of value, or "a future like ours." And, according to Marquis, a future of value is: "that set of experiences you will value in the remainder of your life" if you do not die now (Marquis, 2011, 12). So, Marquis does not first determine which individuals have basic rights and then (as a consequence) determine which future states of affairs are beneficial for those individuals and therefore are objects of rights. Rather, he first determines which states of affairs are valuable, and then concludes that any being that now possesses a future containing such experiences has basic rights, including a right to life.

In short, for Marquis, S is intrinsically valuable only if S has a future that is of value. And a state of affairs X is valuable to S only if S *would actually value it.* As Marquis explains: "The future of value account is based on what has value. It is based on what, in the victim's future, she would value. Therefore, it is superior to those accounts of the wrongness of killing based on some property not essentially connected with value" (Marquis, 2011, 4).

A central problem for this account is that in the end it is based simply on the claim that we would value certain states of affairs but not others. For example, Marquis contends that only a "biographical life," (a life filled with positive experiences), as opposed to a "mere biological life" is something all of us would value. So, if S has a future with a positive *biographical* life (a future life with positive mental experiences), then S has a future of value and therefore S himself or herself is valuable, but not if S has only a "mere biological life." Yet, as Marquis himself notes when criticizing Singer, some individuals do not value their future lives even though their futures include lives with many positive mental experiences—what he would call a positive "biographical" as well as a "mere biological" life. For example, those suffering from depression or brainwashing. Marquis rightly holds that such individuals do have intrinsically valuable lives despite their not actually valuing them. So, his appeal actually needs to be not just to a future that we *do value* but to a future that it is *reasonable* for us to value.

Marquis holds that a person's future of value is "that set of *experiences* you will value in the remainder of your life" if you do not die now (Marquis, 2011, 12, emphasis added). Marquis explicitly eschews body-self dualism, but the terminology—"biographical life" vs. mere "biological life"—suggests that human beings have an *im*personal, merely biological aspect, and a mental aspect that alone is truly personal and inherently valuable. On the contrary, while human beings are animals, insofar as they are animals with a rational and free nature, they are through-and-through personal and their life is never merely biological.

Marquis notes that proponents of the substantial identity view might argue that what is of value are *real goods* as opposed to experiences. As he notes, the difference between the two positions about what is worthwhile pursuing is that the "goods account" allows that there may be real goods or benefits that lack experiential content. On the goods account, Marquis indicates, one might hold that biological life itself, even apart from experiences, is a good, worthy of pursuit. Marquis argues that, while bodily life does have important *instrumental* value, this does not show it has intrinsic value (nor, he concedes, that it does *not* have intrinsic value). According to Marquis, the claim that mere human biological life has intrinsic value needs an argument to support it, though he himself offers no argument for his *denial* of that claim (besides the speculation that most people would not want to be preserved in a vegetative state). He asserts that, "For myself, I cannot imagine why I would care if my life were ended if I were to fall into an irreversible vegetative state" (Marquis, 2011, 18).

However, it seems that, having distinguished the "experientialist" view of what has value from the genuine goods view, or real fulfillment view, Marquis simply re-asserts his experientialist position. He has not given an argument to support that view (nor has he claimed to). I believe, though, that there are good reasons to favor the goods view over the experientialist view of value (Lee and George, 2008a, Chapter 3), and that part and parcel of the goods view is that a human being's bodily life, even if severely debilitated so that one's conscious capacities will not (or likely will not) be actualized, remains inherently valuable.

Marquis does not, I think, see the full import of the "goods version" of the future of value argument. If what makes a condition worth pursuing is that it is really fulfilling for beings with a rational nature, then it is hard to see why biological life would *not* be for them a basic good. Marquis may be right that the denial that biological life is basic good is not *contradictory* to that position. Still, the experientialist view of what has value seems to be the only rationale for that denial.

In previous writings I have argued that the denial that biological life is intrinsically good involves an implicitly dualist view of the human person, an implicit identification of the self with something other than the bodily entity that a human being actually is (a non-material subject, or a set of experiences) (Lee, 1998; Lee and George, 2008a, 130–133). My argument has never been an attempt to deduce an ought-proposition from premises all of which are is-propositions—in my judgment that would be a fallacy involving a logically illicit modal shift. Rather, I have presumed that you and I recognize that what each of us is, or the thing that each of us is, is intrinsically valuable or worth pursuing or promoting. So, if the thing I am is intrinsically valuable, but this bodily entity, this human organism, were not intrinsically valuable, it *would* follow that I am something other than this bodily entity. Marquis rejects this argument:

> No one (except the few dualists among us) doubts that Lee and George are living bodily entities. No one who thinks they are living bodily entities doubts that they are intrinsically valuable, whether they are intrinsically valuable in virtue of their essential properties or their accidental properties. Does it follow that those two bodily entities themselves are intrinsically valuable *independently* of any properties they may possess? One needs an argument here that accidental properties are not sufficient to give one value (Marquis, 2011, 18).

Marquis is right. In this argument, "intrinsically valuable" means: valuable as an end and not merely as a means, and a thing could be valuable as an end but at the same time require an accidental property to be valuable as an end.[19] Compare: I value Jones, Jones is essentially human, therefore as long as Jones is human, I must, by logical consistency, value Jones. That of course does not follow. (This is similar to the opacity of intentional acts: for example, from the fact that I desire to drink that liquid, and that liquid is denatured alcohol, it does not follow that I desire to drink denatured alcohol.) However, if I add that I value Jones because he is honest, or in virtue of his honesty, and he is still honest, then I must still value Jones's honesty, even if he has acquired vices that make him overall repulsive to me. In other words, if what makes him likeable is his honesty, and he is still honest, then at least that feature in him must still be likeable. When we deliberate about what to do, that is, when we reason practically, what makes a condition valuable as an end (intrinsically valuable) is that it is *really fulfilling* for myself (and others like me). And the preservation of my being (and of others like me) is *really fulfilling for me*. Now, if that is so, then to hold that my biological life is *not* good as an end (whether overall the means to preserve it are worth pursuing, given the difficulties of those means) is implicitly to identity the preservation of myself with the preservation of something other than this bodily entity.

Two further points. First, this means that if it *is true* that what makes a condition valuable is its being really fulfilling (for myself and/or others like me) then one who denies that one's biological life is good as an end is implicitly identifying himself with something other than the biological entity he is, whether he explicitly acknowledges that what makes something good as an end is its being really fulfilling or not. Second, this argument nevertheless will not be convincing except to those who do explicitly hold that what makes a condition good as an end is its being really fulfilling for myself (and others like me, however that likeness is parsed out).

Marquis replies to something I have said that is very close to this argument. I had said that it makes no sense to value the fulfillment of a thing (a human being) but not the thing itself. "It makes no sense to hold that the fulfillment of an entity is intrinsically valuable, and yet the entity itself is not" (Lee and George, 2008a, 161). But Marquis replies that even though it is true that the fulfillments of me do include me, "it does not follow that if I am not (indeed, cannot be) fulfilled by experiences that are good for me that I have any value at all" (Marquis, 2011, 18). Again, Marquis has a point: there is no contradiction in holding that the fulfillment of X is valuable but X is not valuable apart from that fulfillment. But the argument was that *what makes something valuable as an end* is that it is really fulfilling for myself (and perhaps others). And from that it *does* follow that the thing itself is valuable, that the preservation of a thing is part of its fulfillment. Marquis (and others) may not be convinced that the feature that makes a condition worth pursuing for a person is his or her (and perhaps others') real fulfillment. Nevertheless, these arguments suggest that Marquis's position—to the extent it differs from the substantial identity account—is tied to what Marquis calls the experientialist version of the future of value account as opposed to the goods version. Since there are good reasons to prefer the goods version over the experientialist version of value, those also are reasons for rejecting Marquis's version of the value of human life.

2.6 Persons in Persistent Vegetative State, and Anencephalic Infants

The difference between the substantial identity view and Marquis's view has concrete applications with respect to those individuals judged to be irreversibly unconscious. On Marquis's view human beings in a persistent vegetative state—if they are judged to be permanently such—and anencephalic babies are not intrinsically valuable because they lack, according to him, a future of value. But on the substantial identity account these human beings still have intrinsic value because they still are individuals with a rational nature.

What has value, what is to be respected, is the individual himself or herself. In other words, the individual is not a mere vehicle for what has value, but is in himself or herself valuable as an end (otherwise, it would be permissible at times to kill one child on condition that we replace him or her with two other, more productive, vehicles of what has value). This brings us to some puzzles. To harm someone is to deprive him of a future he might otherwise have had. And, as Marquis points out in his paper, the basis in reality for our knowledge of the future an individual would have had is (at least in part) his or her essential nature. However, suppose S is about to be killed by someone else, but I kill him first. Can I reason that since S was about to be killed by someone else anyway, I have not deprived him of a future of value, or at least not very much of a future of value (say, minutes or seconds)? I don't think so. The fact that another person is about to kill S does not diminish the basic moral wrongness of my killing S.

Now, suppose that what is about to kill S is not another person, but a non-rational cause, such as a bacterium or a virus? The basic wrongness would remain. But it also would remain if the cause about to kill him was an internal *defect*, such as a genetic defect, a brain hemorrhage, atherosclerosis, or so on.[20] So, the fact that a person is about to die anyway does not remove or substantially diminish the basic wrongness of my killing him: the fact that *I* would be the deliberate cause of his death is decisive in what makes killing someone wrong, not just whether my action makes a difference in the overall future state of the world.[21] So, unlike Marquis, I do not think the basic moral wrongness of killing a person varies with respect to that individual's life expectancy, or the quality of life that is expected for him or her. (It may be *aggravated* or *mitigated* by such factors, but the essential grave moral wrong of the intentional killing of an innocent person will remain.) If that is so, then in the case of an anencephalic child, the fact that he or she is about to die does not mean it is morally right to kill this child.

Moreover, the fact that there may be defects impeding this individual from realizing that to which his or her nature orients this individual, also does not make killing him or her morally right. Just as the fact that something else might be causing (or about to cause) an individual's death does not justify my killing him, so the fact that something else might be impeding a person's development or actualization does not justify my impeding his development. Thus, a human individual remains intrinsically valuable despite imminent causes of his death and/or despite impediments (whether external or internal) to his development. So, a human being in a persistent vegetative state is intrinsically valuable as a subject of rights despite the fact that internal factors impede his or her development. Intentionally killing such a person is gravely morally wrong.[22] So, every human being, including those in persistent vegetative state and anencephalic infants, are persons and intrinsically valuable as subjects of rights.

Notes

1. For example, Boyle, 1979; along with others, I also have defended this view: Lee, 1996, 2004; Lee and George, 2008a, Chapter 4; see also Grisez, 1990.
2. Hence this position is not a "speciesist" view.
3. See for example Singer, 1999, Chapters 4 and 6; Dworkin, 1993; and Steinbock, 1992.
4. Marquis, 2011. I wish to thank Don Marquis for this very incisive and challenging article.
5. We set aside the issues of capital punishment and war—and so "human being" should be read as an innocent human being, that is, a human being who is not an enemy combatant and has not been convicted of a capital crime.

6. It is important to note that having a basic natural capacity is not an accidental characteristic in entities that have it: it indicates the basic kind of being they are. So one cannot lose that basic natural capacity and survive; for example, if the materials in my makeup cease to form a rational animal, then I cease to be, that is, I have died. See below, note 29.
7. A serious problem with the desire or interest account is that it seems to imply that even human beings in temporary comas are not persons, lack rights, and therefore it would be morally permissible to kill them. Proponents of the desire or interest account might reply that the individual's *past* desires ground his or her rights. But this reply fails. Suppose I had a type of brain surgery that caused a somewhat prolonged coma, followed by permanent amnesia. After I awoke I could gradually re-learn my language skills, basic knowledge of arithmetic, history, etc., but I would never regain knowledge of my past life, including my past relationships, plans, projects, and desires. While I was in a coma I would surely have a right to life; it would be wrong intentionally to kill me. But I would not possess any present desires, and, since I would never regain my previous desires, memories, and so on, this right could not be based on my *past* desires either. So past desires do not ground the right to life of people in coma. For an argument along these lines, see Beckwith, 2007, 130–140.Or a proponent of the desire or interest account might modify the position so that what grounds rights are *ideal* desires—what an individual *would* desire if he or she were fully informed and rational (see Boonin, 2003). But, as Marquis has pointed out, this account would not exclude human embryos and fetuses from having rights: an embryo or fetus *would* desire to continue to live, if he or she did not have the handicap of underdevelopment. If one modifies the ideal desire account to say that the idealization must be of desires one actually has, and thus human embryos and fetuses are excluded (Boonin, 2003), the move seems *ad hoc;* moreover, on this view someone in a coma who will regain consciousness, but not the same conscious experiences (memories, skills, etc.) he once had, would—implausibly—not have a right to life. Then too, to determine what counts as an ideal desire one needs a standard, and, arguably, an objective standard rooted in the agent's real fulfillment.
8. I hold we should *never* intentionally kill entities that possess basic rights, but that need not be discussed here.
9. It is worth noting that natural kinds are classified by their basic natural capacities, not their immediately exercisable ones: kittens for example are classified as carnivores.
10. Note that the argument is not just that there is no radical difference between any two *adjacent* stages of development of a human being, but that the difference between a human being at one stage of development and that same human being at *any* other stage of development is only one of degree. Hence the objection that this argument commits the *sorites* fallacy, as is sometimes claimed, is mistaken. See, for example: Sandel, 2007, 118. A more detailed reply to Sandel's objection will be found in George and Lee, 2009.

11. "Can this be the basic natural capacity to exhibit rational behavior? I have had that basic natural capacity as long as I have been alive. *That capacity* does not very in degree over the time of its development (although its behavioral manifestations may increase over time)" (Marquis, 2011, 9).
12. In previous works I expressed this argument confusedly. See, for example, Lee, 2004, where I begin by saying that there are two types of capacities, basic natural capacities and immediately exercisable capacities, and then argue that the criterion for being a subject of rights should be former not the latter. But later in this article I say, "The proximate, or nearly immediately exercisable capacity for mental functions is only the development of an underlying potentiality that the human being possesses simply by virtue of the kind of entity it is." Actually, the latter point is the central one, and negates the first claim, or at least relegates it to a fact mainly about language.
13. Of course, we do form a distinct *concept* of *an immediately exercisable capacity*, and it is distinct from the concept of a natural, basic capacity and we have some criteria for the application of this concept or predicate to different phases in an organism. The point, however, is that the distinct concepts do not correspond to really distinct properties.
14. Thanks to Don Marquis for suggesting this objection by email.
15. I distinguish actual consciousness from the disposition to have actual consciousness. David Boonin (Boonin, 2003) proposes to base the right to life on actual desires in the sense of dispositional desires, that is, dispositions to have desires. However, if one does not qualify the kind of disposition for desires and consciousness an entity must have to have a right to life, the result will be that human embryos and fetuses do have a right to life since they surely have—in some real sense—such dispositions. If one qualifies the kind of disposition needed with the result that human embryos and fetuses are excluded, then the move will seem *ad hoc;* moreover, it will be hard to see how someone in an amnesia coma would have a right to life.
16. Also, one might object that there is a qualitative difference between the potentiality to develop oneself to the point where one is conscious, on the one hand, and the stage at which one actually has all of the internal resources needed to be conscious, lacking only suitable external circumstances needed for its actuation, on the other hand. And one might then label the latter stage an immediately exercisable capacity for consciousness, and make that the criterion for possession of basic rights. As I indicated above in explaining Singer's position, there is a difference between these two types of states, but they are not qualitatively different; the one is in reality just a further actuation of the other. Michael Tooley distinguishes between the two, and calls the former stage a "potentiality" for consciousness and the latter an "immediately exercisable capacity" for consciousness, and says that only the latter is ethically significant. But, to maintain this, Tooley must also distinguish between an immediately exercisable capacity for consciousness that is *blocked,* and one that is not. A blocked capacity, according to Tooley, is one that has only "negative factors" preventing

it from actualization. So it still qualifies—according to Tooley—as an immediately exercisable capacity. Immediately exercisable capacities are ones that do not need to undergo internal change before they are exercised (Tooley, 1983, 146 ff.).

However, there is no qualitative difference between factors that merely block the actualization of a capacity, on the one hand, and those that introduce an internal change into the organism so that it is (once again, perhaps) at a stage where it needs only the suitable external circumstances to exercise its capacity, on the other hand. For example, an organism might be able to see (have an immediately exercisable capacity to see) but be impeded from doing so because it is in pitch darkness (an external impediment). But it might also be impeded from seeing because it is anesthetized. And recovering from anesthesia involves *internal* changes. Hence there is no significant difference between the kind of capacity for self-consciousness possessed by a human embryo, fetus, or infant, and the kind possessed by an anesthetized human adult. Both must internally change, in addition to being placed in suitable circumstances, in order to acquire consciousness. Indeed, unconsciousness in adults is due to the same type of condition that prevents consciousness in a human fetus or infant with an immature brain, namely, a lack of electrical connections between different parts of the brain.

17. In previous writings by "basic natural capacity" I meant to refer to the substantial nature itself. This point is made in Lee, 2007.
18. I suppose one might say that I am treating *having a rational nature* as a threshold property. But that is not the case: what is called a threshold property is actually the possession of some degree of another property, where having that degree abruptly causes another effect (for example, the threshold stimulus reached in a neuron), or is treated by general agreement as a cut-off point for various purposes (as in intoxication). On the substantial identity view, a subject that is inherently oriented to a rational mode of life is different in kind, not just in degree, from non-rational individuals, and every rational being, qua rational being, is equal.
19. Dean Stretton first made this clear to me in Stretton, 2004.
20. Suppose he's about to commit suicide—would my killing him be permissible? If not, this shows that the defect's being internal does not matter for the wrongness of the deliberate killing even if it might mitigate the seriousness of that wrong.
21. See Bernard Williams's argument in Smart and Williams, 1975.
22. Marquis also argues that human beings who have been accurately diagnosed with brain death are nevertheless still human beings. But (he argues) they lack the basic natural capacity for rationality and so this shows that a premise in one of our arguments is false (namely: Every human individual has a rational nature.). On this point Marquis follows D. Alan Shewmon; see for example: Shewmon, 2001. However, my view now is that complete brain death *is* the death of the human being, because even if there is in some sense a single individual alive after brain death (which Marquis, following Shewmon, points out

does occur in some cases), that is not the same individual as was alive before, and it is not a human individual, because it no longer has the basic natural capacity for sentience or (as a consequence) for rationality. In other words, if there is a single organism alive after brain death, it is not a rational animal. So, a substantial change has occurred. The individual that was alive has died, and what is alive is not a human being.

If after decapitation of a human being both the head and the headless body could be kept alive for a short time (by suturing the blood vessels, supplying a heart-lung machine for the head, and a ventilator for the headless body), the two entities obviously could not both be the same individual who lived before decapitation. I also think the headless body would *not* be a human being, even if some interior organic integration emerged among its bodily organs so that it had a type of internal unity. The reason why is that it would not be a rational animal. It would not have the capacity for sentience, nor the basic natural capacity for sentience (it would have neither a brain *nor* the capacity to develop a brain—the human embryo has that, which is why he or she is an animal and given the sort of brain it has the capacity to develop, a rational animal) and so it would not be a sentient being, that is, an animal. And so it would not be a rational animal, a human being. On this point, see Lee and Grisez (2011).

References

Beckwith, F.P. 2007. *Defending life: A moral and legal case against abortion choice*. Cambridge, MA: Cambridge University Press.

Boonin, D. 2003. *A defense of abortion*. New York, NY: Cambridge University Press.

Boyle, J.M. 1979. That the fetus should be considered a legal person. *The American Journal of Jurisprudence* 59:59–71.

Dworkin, R. 1993. *Life's dominion: An argument about abortion, euthanasia, and individual freedom*. New York, NY: Alfred A. Knopf.

George, R.P., and P. Lee. 2009. Embryonic human persons. Talking point on morality and human embryo research. *EMBO Reports* 10:301–306.

Grisez, G. 1990. When do people begin? *Proceedings of the American Catholic Philosophical Association* 63:27–47.

Lee, P. 1996. *Abortion and unborn human life*. Washington, DC: Catholic University of America Press.

Lee, P. 1998. Human beings are animals. In *Natural Law and Moral Inquiry, Ethics, Metaphysics, and Politics in the Work of Germain Grisez*, ed. Robert P. George, 133–152. Washington, DC: Georgetown University Press.

Lee, P. 2004. The pro-life argument from substantial identity: A defense. *Bioethics* 18:249–263.

Lee, P. 2007. Substantial identity and the right to life: A rejoinder to Dean Stretton. *Bioethics* 21:93–97.

Lee, P., and R.P. George. 2008a. *Body-self dualism in contemporary ethics and politics*. Cambridge, MA: Cambridge University Press.

Lee, P., and R.P. George. 2008b. The nature and basis of human dignity. In *Human dignity and bioethics, essays commissioned by the president's council on bioethics*, 409–434. Washington, DC, reprinted in *Ratio Juris*, 21 (2008), 173–193.

Lee, P., and G. Grisez. 2011. *Bioethics*. Total brain death: A reply to Alan Shewmon.

Marquis, D. 1989. Why abortion is immoral. *The Journal of Philosophy* 86:138–202.

Marquis, D. 2011. Why abortion is seriously wrong: Two views. In *Bioethics with liberty and justice: Themes in the work of Joseph Boyle*, ed. C. Tollefsen, 3–22. Dordrecht, The Netherlands: Springer.

Sandel, M. 2007. *The case against perfection*. Cambridge, MA: Belknap Press of Harvard University.

Shewmon, D.A. 2001. The brain and somatic integration: Insights into the standard biological rationale for equating 'Brain Death' with death. *Journal of Medicine and Philosophy* 26: 457–478.

Singer, P. 1999. *Practical ethics, 2nd edition*. New York, NY: Cambridge University Press.

Smart, J.J.C., and Bernard Williams. 1975. *Utilitarianism: For and against*. New York, NY: Cambridge University Press.

Steinbock, B. 1992. *Life before birth*. New York, NY: Oxford University Press.

Stretton, D. 2004. Essential properties and the right to life: A response to Lee. *Bioethics* 18: 254–281.

Tooley, M. 1983. *Abortion and infanticide*. New York, NY: Oxford University Press.

Part II
Moral and Legal Issues at the Beginning and Ending of Life

Chapter 3
Embryo Ethics: Justice and Nascent Human Life

Robert P. George

3.1 Introduction

If we were to contemplate killing mentally handicapped infants to obtain transplantable organs, no one would characterize the controversy that would erupt as a debate about organ transplantation. The dispute would be about the ethics of killing handicapped children to harvest their vital organs. We could not resolve the issue by considering how many gravely ill people we could save by extracting a heart, two kidneys, a liver, etc., from each mentally handicapped child. Instead, we would have to answer this question: is it right to relegate a certain class of human beings—the handicapped—to the status of objects that can be killed and dissected to benefit others?

By the same token, strictly speaking ours is not a debate about stem cell research. No one would object to the use of pluripotent stem cells in biomedical research or therapy if they could be obtained from non-embryonic sources, or if they could be acquired by using embryos lost in miscarriages.[1] The point of controversy is the ethics of deliberately destroying human embryos to produce stem cells. The threshold question is whether it is right to kill members of a certain class of human beings–those in the embryonic stage of development–to benefit others.

Supporters of embryo-destructive research insist, however, that human embryos are not human beings—or if they are human beings, that they are not yet "persons." It is therefore morally acceptable, they say, to "disaggregate" them for the sake of research aimed at finding cures or treatments for juvenile diabetes and other horrible afflictions.

At the heart of the debate over embryo-destructive research, then, are two questions: is a human embryo a human being, and, if so, what is owed to an embryonic human as a matter of justice? Such questions have consistently been a matter of philosophical concern to Joseph Boyle, and he has answered them with characteristic philosophic rigor.[2] The arguments that follow continue work that he, with

R.P. George (✉)
Princeton University, Princeton, NJ, USA
e-mail: rgeorge@princeton.edu

C. Tollefsen (ed.), *Bioethics with Liberty and Justice*, Philosophy and Medicine 110,
DOI 10.1007/978-90-481-9791-0_3, © Springer Science+Business Media B.V. 2011

Germain Grisez, John Finnis, and others, have carried out in defense of justice for the youngest members of our species.

3.2 Framework of the Discussion

I will say nothing about religion or theology. This is not a tactical decision; rather, it reflects my view about how to think about the dispute over killing human embryos. It is sometimes said that opposition to embryo-destructive research is based on a controversial theology of "ensoulment."

But one need not engage questions of whether human beings have spiritual souls in considering whether human embryos are human beings. Nor must one appeal to any theology of ensoulment to show that there is a rational basis for treating all human beings–including those at the embryonic stage–as creatures possessing intrinsic worth and dignity.[3]

We should resolve our national debate over embryo-destructive research on the basis of the best scientific evidence as to when the life of a new human being begins, and the most careful philosophical reasoning as to what is owed to a human being at any stage of development. Religious conviction can motivate us to stand up and speak out in defense of human life and dignity. And religious people should never hesitate to do that. But we need not rely on religious authority to tell us whether a human embryo is a new living member of the species *Homo sapiens* or whether all human beings–irrespective of not only race, ethnicity, and sex but also age, size, stage of development, and condition of dependency–possess full moral worth and dignity. The application of philosophical principles in light of facts established by modern embryological science is more than sufficient for that task.[4]

3.3 What the Human Embryo Is

The adult human being that is now you or me is the same human being who, at an earlier stage, was an adolescent and, before that, a child, an infant, a fetus, and an embryo.[5] Even in the embryonic stage, you and I were undeniably whole living members of the species *Homo sapiens*. We were then, as we are now, distinct and complete—though, in the beginning, developmentally immature–human organisms. We were not mere parts of other organisms.

A human embryo is not something different *in kind* from a human being, like a rock, or a potato, or a rhinoceros. A human embryo is a human individual in the earliest stage of his or her natural development.[6] Unless severely damaged or deprived of a suitable environment, an embryonic human being will, by directing his or her own integral organic functioning, develop himself or herself to each new stage of developmental maturity along the gapless continuum of a human life. The embryonic, fetal, infant, child, and adolescent stages are just that: *stages* in the development of a determinate and enduring entity—a human being—who comes into existence as a single-celled organism (zygote) and grows, if all goes well, into adulthood many years later.[7]

By contrast, the gametes whose union brings into existence the embryo are not whole or distinct organisms. Each is functionally (and genetically) identifiable as *part* of the male or female (potential) parent. Moreover, each gamete has only half the genetic material needed to guide the development of an immature human being toward full maturity. They are destined either to combine with an oocyte or spermatozoon and generate a new and distinct organism, or simply to die. When fertilization occurs, they do not survive; rather, their genetic material enters into the composition of a new organism.

But none of this is true of the human embryo, from the zygote and blastula stages onward. The combining of the chromosomes of the spermatozoon and of the oocyte generates what human embryology identifies as a new, distinct, and enduring organism. Whether produced by fertilization, Somatic Cell Nuclear Transfer (SCNT), or some other cloning technique, the human embryo possesses all of the genetic material and non-genetic qualities needed to inform and organize its growth.[8] The direction of its growth is not extrinsically determined, but is in accord with the information *within* it.[9] Nor does it merely possess organizational information for maturation; rather, it has an *active disposition* to develop itself using that information. The human embryo, then, is a whole and distinct human organism—an embryonic human being.

If the embryo were not a complete organism, what could it be? Unlike the spermatozoa and the oocytes, it is not merely a part of a larger organism, namely, the mother or the father. Nor is it a disordered growth or gamete tumor, such as a complete hydatidiform mole or teratoma.

Someone might say that the early embryo is an intermediate form, something which regularly emerges into a whole human organism but is not one yet. But what could cause the emergence of the whole human organism, and cause it with regularity? As I have already observed, from the zygote stage forward the development of this organism is *directed from within*, or by the organism itself. So, after the embryo comes into being, no event or series of events occur that we could construe as the production of a new organism–that is, nothing extrinsic to the developing organism itself acts on it to produce a new character or a new direction in development.[10]

3.4 Are All Human Beings Owed Full Moral Respect?

A supporter of embryo-destructive research might concede that a human embryo is a human being in a biological sense, yet deny that we owe human beings in the early stages of their development full moral respect, such that we may not kill them to benefit more fully developed human beings who are suffering from afflictions.

But to say that embryonic human beings do not deserve full respect, one must suppose that not every human being deserves full respect. And to do that, one must hold that those human beings who warrant full respect deserve it not by virtue of the *kind of entity they are*, but, rather, because of some acquired characteristic that some human beings (or human beings at some stages) have and others do not, and which some human beings have in greater degree than others do.

This position is untenable. One need not be *actually* or immediately conscious, reasoning, deliberating, making choices, etc., in order to be a human being who deserves full moral respect, for plainly we should accord people who are asleep or in reversible comas such respect. But if one *denied* that human beings are valuable by virtue of what they are, and required an additional attribute, it would have to be a capacity of some sort, and, obviously, a capacity for certain mental functions.

Of course, human beings in the embryonic, fetal, and early infant stages lack immediately exercisable capacities for mental functions characteristically carried out by most human beings at later stages of maturity. Still, they possess these very capacities *in principe vel radice,* that is, in radical or "root" form. Precisely by virtue of the kind of entity they are, they are from the beginning actively developing themselves to the stages at which these capacities will (if all goes well) be immediately exercisable. Although, like infants, they have not yet developed themselves to the stage at which they can perform intellectual operations, it is clear that they are *rational animal organisms.*[11] That is the *kind* of entity they are.

3.5 Defining Capacities

Here, it is important to distinguish two senses of the "capacity" for mental functions: an immediately exercisable capacity, and a basic natural capacity, which develops over time. We have good reason to believe that the second sense, and not the first, provides the basis for regarding human beings as ends in themselves, and not as means only—as subjects possessing dignity and human rights, and not as mere objects.

First, the developing human being does not reach a level of maturity at which he or she performs a type of mental act that other animals do not perform—even animals such as dogs and cats—until at least several months after birth. A six-week-old baby lacks the *immediately exercisable* capacity to form abstract concepts, engage in deliberation, and perform other characteristically human mental functions. If we owed full moral respect only to those who possess immediately exercisable capacities for characteristically human mental functions, it would follow that six-week-old infants do not deserve full moral respect.[12] Therefore, if we may legitimately destroy human embryos to advance biomedical science, then logically, subject to parental approval, the body parts of human infants should also be fair game for scientific experimentation.[13]

Second, the difference between these two types of capacity is merely a difference between stages along a continuum. The immediately exercisable capacity for mental functions is only the development of an *underlying* potentiality that the human being possesses simply by virtue of the kind of entity it is. The capacities for reasoning, deliberating, and making choices are gradually brought toward maturation, through gestation, childhood, adolescence, and so on. But the difference between a being that deserves full moral respect and a being that does not (and can therefore legitimately be killed to benefit others) cannot consist only in the fact that while both have some feature, one has *more* of it than the other. A mere *quantitative* difference cannot by itself provide a justification for treating entities in radically different ways.[14]

Third, the acquired qualities proposed as criteria for personhood, such as self-consciousness or rationality, come in an infinite number of degrees. If human beings are worthy of full moral respect only because of such qualities, and those qualities come in varying degrees, humans should possess rights in varying degrees. The proposition that all human beings are created equal would be relegated to the status of a myth: since some people are more rational than others (that is, have developed that capacity to a greater extent than others have), some people would be greater in dignity than others, and the rights of the superiors would trump those of the inferiors.[15]

So it cannot be the case that some human beings and not others are intrinsically valuable by virtue of a certain degree of development. Rather, *all* human beings are intrinsically valuable (in the way that enables us to ascribe to them equality and basic rights) because of the *kind* of being they are.

Since human beings are intrinsically valuable and deserve full moral respect by virtue of *what* they are, it follows that they are intrinsically and equally valuable *from the point at which they come into being*. Even in the embryonic stage of our lives, each of us was a human being and, as such, worthy of concern and protection. Embryonic human beings, whether brought into existence by union of gametes, SCNT, or other cloning technologies, should be accorded the respect given to human beings in other developmental stages.[16]

3.6 Somatic Cells and Embryos

I wish to turn now to some arguments that advocates of embryo-destructive research have advanced to cast doubt on the proposition that human embryos deserve to be accorded full moral status.

In defending research involving the destruction of human embryos, Ronald Bailey, a science writer for *Reason* magazine, developed an analogy between embryos and somatic cells in light of the possibility of human cloning (Bailey, 2001). Bailey claims that every cell in the human body has as much potential for development as any human embryo. Embryos therefore have no greater dignity or higher moral status than ordinary somatic cells. Bailey observes that each cell in the human body possesses the entire DNA code; each has become specialized (as muscle, skin, etc.) because most of that code has been turned *off*. In cloning, those previously deactivated portions of the code are reactivated. So, Bailey says, quoting Australian bioethicist Julian Savulescu, "if all our cells could be persons, then we cannot appeal to the fact that an embryo could be a person to justify the special treatment we give it." Since plainly we are not prepared to regard all of our cells as human beings, we should not regard embryos as human beings.

Bailey's analogy between somatic cells and human embryos collapses, however, under scrutiny. The somatic cell is something from which (together with extrinsic causes) a new organism can be *generated* by the process of somatic cell nuclear transfer, or cloning; it is certainly not, however, a distinct organism. A human embryo, by contrast, already is a distinct, self-developing, complete human organism.

Bailey suggests that the somatic cell and the embryo are on the same level because both have the "potential" to develop to a mature human being. The kind of "potentiality" possessed by somatic cells that might be used in cloning differs profoundly, however, from the potentiality of the embryo. A somatic cell has a potential only in the sense that something can be done to it (or done with it) so that its constituents (its DNA molecules) enter into a distinct whole human organism, which is a human being, a person. In the case of the embryo, by contrast, he or she already is actively—indeed dynamically–developing himself or herself to the further stages of maturity of the distinct organism–the human being—he or she already is.

True, the whole genetic code is present in each somatic cell; and this code can guide the growth of a new entire organism. But this point does nothing to show that a somatic cell's potentiality is the same as a human embryo's. When scientists remove the nucleus of an ovum, insert the nucleus of a somatic cell into the remainder of the ovum, and give it an electric stimulus, they are doing more than merely placing the somatic cell in an environment hospitable to its continuing maturation and development. They are generating a wholly distinct, self-integrating, entirely new organism—an embryo, in other words. The entity—the embryo–brought into being by this process is radically different from the constituents that entered into its generation.

Somatic cells, in the context of cloning, then, are analogous not to embryos, but to the gametes whose union results in the generation of an embryo in the case of ordinary sexual reproduction. You and I were never either a sperm cell or an ovum. Nor would a person who was brought into being by cloning have once been a somatic cell. To destroy an ovum or a skin cell whose constituents might have been used to generate a new and distinct human organism is not to destroy a new and distinct human organism—for no such organism exists or ever existed. But to destroy a human embryo is precisely to destroy a new, distinct, and complete human organism—an embryonic human being.[17]

3.7 The Brain Death Argument

Michael Gazzaniga, a psychologist and neuroscientist at the University of California, Santa Barbara, has proposed a different argument. While agreeing that a human embryo is an entity possessing a human genome, he has suggested that a "person" comes into being only with the development of a brain. Prior to that point we have a human organism, but one lacking the dignity and rights of a person (President's Council on Bioethics, 2002). We may therefore legitimately treat human beings in the earliest stages of development as we would treat organs available for transplantation (assuming, as with transplantable organs, that proper consent for their use is given, etc.).

In presenting his case, Gazzaniga observes that modern medicine treats the death of the brain as the death of the person—authorizing the harvesting of organs from the remains of the person, even if some physical systems are still functioning. If a human being is no longer a person with rights once the brain has died, then surely a human being is not yet a person prior to the development of the brain.

This argument suffers, however, from a damning defect. Under prevailing law and medical practice, the rationale for brain death is not that a brain-dead body is a living human organism but no longer a person. Rather, brain death is accepted because the irreversible collapse of the brain is believed to destroy the capacity for self-directed integral organic functioning in human beings who have matured to the stage at which the brain performs a key role in integrating the organism. In other words, at brain death a unitary organism is believed no longer to exist.[18] By contrast, although an embryo has not yet developed a brain, it is clearly exercising self-directed integral organic functioning, and so it *is* a unitary organism. Its capacity to develop a brain is inherent and progressing, just as the capacity of an infant to develop its brain sufficiently for it actually to *think* is also intrinsic and unfolding.

Unlike a corpse—the remains of what was once a human organism but is now dead, even if particular systems may be artificially sustained—a human organism in the embryonic stage of development is a complete, unified, self-integrating human individual. It is not dead but very much alive, even though its self-integration and organic functioning are not brain-directed at this stage. Its future lies ahead of it, unless it is cut off or not permitted to develop its inherent capacities. Therefore, defenders of embryonic human life insist that the embryo is not a "potential life," but is rather a life *with potential*. It is a potential *adult*, in the same way that fetuses, infants, children, and adolescents are potential adults. It has the potential for agency, just as fetuses, infants, and small children do. Just like human beings in the fetal, infant, child, and adolescent stages, human beings in the embryonic stage are already, and not merely potentially, *human beings*.[19]

3.8 Acorns and Embryos

In an essay in the *New England Journal of Medicine*, Harvard political theorist Michael Sandel claimed that human embryos are different *in kind* from human beings at later developmental stages. This argument truly takes us to the heart of the matter: is a human embryo a human being? At its core is this analogy:

> Although every oak tree was once an acorn, it does not follow that acorns are oak trees, or that I should treat the loss of an acorn eaten by a squirrel in my front yard as the same kind of loss as the death of an oak tree felled by a storm. Despite their developmental continuity, acorns and oak trees are different kinds of things (Sandel, 2004, 208).

He maintains that just as acorns are not oak trees, embryos are not human beings.

Sandel's argument begins to go awry with his choice of analogates. The acorn is analogous to the embryo, and the oak tree (he says) is analogous to the human being. But in view of the developmental continuity that science fully establishes and Sandel concedes, the proper analogate of the oak tree is the *mature* human being, viz., the adult. Sandel's analogy has its apparent force because we feel a sense of loss when a mature oak is felled—assuming it is a magnificent or beautiful oak. But while it is true that we do not feel the same sense of loss at the destruction of an acorn, it is also true that we do not feel the same sense of loss at the destruction of an oak *sapling*. But clearly the oak tree does not differ in kind from the oak sapling.

This example shows that we value oak trees not because of the kind of entity they are, but because of their magnificence. The magnificence of an oak tree reflects either accidental properties or instrumental worth; a mature tree provides our house with shade that is aesthetically pleasing to behold. Neither acorns nor saplings are magnificent, so we do not experience a sense of loss when they are destroyed. If oak trees were valuable by virtue of the *kind* of entity they are, then it would follow that it is just as unfortunate to lose an acorn as an oak tree.

But the basis for our valuing human beings is profoundly different from the basis for valuing oak trees. As Sandel concedes, we value human beings precisely because of the *kind* of entities they are. Indeed, that is why we consider all human beings to be equal in basic dignity and human rights. We most certainly do not believe that especially magnificent human beings—such as Michael Jordan or Albert Einstein—are of greater *fundamental* worth and dignity than human beings who are physically frail or mentally impaired. We would not tolerate the killing of a handicapped child or a person suffering from, say, brain cancer in order to harvest transplantable organs to save Jordan or Einstein.

And we do not stand for the killing of infants, *which on Sandel's analogy would be precisely analogous to the oak saplings whose destruction we do not necessarily regret*. Managers of oak forests freely kill saplings, just as they might destroy acorns, to ensure the health of the more mature trees. No one gives it a second thought. This is precisely because we do not value members of the oak species—as we value human beings—because of the *kind* of entity they are. If we did value oaks in this way, then we would have no less reason to regret the destruction of saplings, and possibly even acorns, than that of mature oak trees. Conversely, if we valued human beings in a way analogous to that in which we value oak trees, then we would have no grounds to object to killing human infants or even mature human beings who were "defective."

Sandel's defense of human embryo-killing on the basis of an analogy between embryos and acorns collapses the moment one brings into focus the profound difference between the basis on which we value oak trees, and that on which we ascribe value to human beings. We value oaks for their accidental properties and their instrumental worth. But we value human beings because of the intrinsic worth and dignity they possess by virtue of the kind of entity they are.[20]

3.9 The Problem of Twinning

I now consider a final objection. Some have claimed that the phenomenon of monozygotic twinning shows that the embryo in the first several days of its gestation is not a human individual. The suggestion is that as long as twinning can occur what exists is not yet a unitary human being, but only a mass of cells–each cell being totipotent and allegedly independent of the others.

It is true that if a cell or group of cells is detached from the whole at an early stage of embryonic development, the detached part can become an organism with the potential to develop to maturity as distinct from the embryo from which it was detached. But this does nothing to show that before detachment the cells within the human embryo constituted only an incidental mass.[21]

Consider the parallel case (discussed by Aristotle) of the division of a flatworm. Parts of a flatworm have the potential to become a whole flatworm when isolated from the present whole of which they are a part. Yet no one would suggest that prior to the division of a flatworm, the original flatworm was not a unitary individual. Likewise, at the early stages of human embryonic development, before specialization by the cells has progressed very far, cells or groups of cells can become whole organisms if they are divided and exist in an appropriate environment after the division. But that fact does not in the least indicate that prior to the twinning event, the embryo is other than a unitary, self-integrating, actively developing human organism. It certainly does not show that the embryo is a mere "clump of cells."

Based on detailed studies of other mammals, it is highly likely that in the first two weeks, the cells of the developing embryonic human being already manifest a degree of specialization and differentiation. From the beginning, even at the two-celled stage, the cells of mouse embryos differ in their developmental fates; they will ultimately contribute to distinct tissues within the embryo.[22] By the four celled stage, there are clear molecular (Torres-Padilla et al., 2007; Stanton et al., 2003) and developmental (Piotrowska-Nitsche et al., 2005) differences between cells of the developing mouse. At no time is the embryo a mere "ball of cells," i.e. a collection of homogeneous cells that do not function together as an organismic whole.

Now some people have claimed that the human embryo does not become a human being until implantation, because (they assume) the embryo cannot establish a basic body plan until it receives external maternal signals at implantation. Only then is it a self-directing human organism. According to this view, these signaling factors somehow transform what was hitherto a mere bundle of cells into a unitary human organism.

However, embryologists argue about whether any such maternal signaling actually occurs. As Hans-Werner Denker observed, it was once assumed that in mammals, in contrast to amphibians and birds, polarity in the early embryo depends upon some external signal, since no clear indications of bilateral symmetry had been found in oocytes, zygotes, or early blastocysts. But this view has been revised in the light of emerging evidence: "[I]ndications have been found that in mammals the axis of bilateral symmetry is indeed determined (although at first in a labile way) by sperm penetration, as in amphibians. Bilateral symmetry can already be detected in the early blastocyst and is not dependent on implantation" (Denker, 2004).

Denker refers specifically to the work of Magdelena Zernicka-Goetz and her colleagues at Cambridge University, and that of R. L. Gardner at Oxford University, which shows that polarity exists even at the two-celled stage. In contrast, Davor

Solter and Takashi Hiiragi of the Max Planck Institute for Immunobiology in Freiburg argue that in the early embryo (prior to compaction and differentiation into inner cell mass and trophoblast), external factors determine the fate of each cell, rather than an internal polarity (Vogel, 2005). In other words, the issue is not definitively settled. However, whichever of the two is true, it is less than candid for anyone to assert the older view without acknowledging that credible scientists from leading universities have published research contradicting it in major peer-reviewed scientific journals.

Moreover—and here is the most important point—even if it *is* the case that polarity does not emerge until a maternal signal is received at implantation, that would *not* provide any evidence that such a signal transformed a bundle of cells into a unitary, multicellular human organism. Just as the lungs begin to breathe at birth only in response to certain external stimuli, so it would make sense (if the older view is true) that differentiation into the rudiments of the distinct body parts (basic bilateral polarity) would begin only in response to some external stimuli. And this is exactly how embryology texts interpreted such signals, even prior to the publications of Zernicka-Goetz and Gardner and their teams.

There is much evidence that the human embryo is from the first day onward a unitary organism, and never a mere bundle of cells. Development in the embryo is complex and coordinated, including compaction, cavitation, and other activities in which the embryo is preparing itself for implantation.

And here is the clearest evidence that the embryo in the first two weeks is not a mere mass of cells but a unitary organism: if each cell within the embryo before twinning were independent, there would be no reason why each would not develop on its own. Instead, these allegedly independent, non-communicating cells regularly function together to develop into a single, more mature member of the human species. This fact shows that the cells are interacting from the very beginning (even within the zona pellucida, before implantation), restraining them from individually developing as whole organisms and directing each of them to function as a relevant part of a single, whole organism continuous with the zygote. The evidence indicates that the human embryo, from the zygote stage forward, is a unitary human organism.[23]

3.10 Conclusion

Supporters of embryo-destructive research have advanced other arguments against the proposition that human embryos are embryonic human beings bearing basic dignity and full moral worth. I have focused in this essay on the strongest arguments against my position and laid aside the weaker ones, such as those proposing to infer something of moral relevance from the fact that human embryos are tiny and not yet sentient; or from the fact that a high percentage of human embryos are naturally lost early in pregnancy; or from the claim that people typically either do not grieve for the loss of embryos in early miscarriages, or grieve but not as intensely as they do for children who die later in gestation or as infants.

If there is a valid argument to show that human embryos are something other than human beings in the embryonic stage of development, or that embryonic human beings lack the basic dignity and moral worth of human beings in later developmental stages, it is one of the arguments I address here. I have given my reasons for believing that none of these arguments can withstand critical scrutiny.

The debate about stem cell research and the value of embryonic human life is sure to continue. But if that debate is informed by serious attention to the facts of embryogenesis and early human development, and of the profound, inherent, and equal dignity of human beings, then we, as a nation, will ultimately reject the deliberate taking of embryonic human life, regardless of the promised benefits.

This does not mean we must sacrifice such benefits altogether. Scientists have already made tremendous progress toward the goal of producing fully pluripotent stem cells by non-embryo-destructive methods. If such methods are pursued with the vigor, the future might see the promise of stem cell science fulfilled, with no stain on our national conscience.[24]

Notes

1. It appears that we will soon be able to obtain embryonic stem cells, or their equivalent, by means that do not require the destruction of human embryos. Important successes in producing pluripotent stem cell lines by reprogramming (or "de-differentiating") human somatic cells have been reported in highly publicized papers by James A. Thomson's research group (Yu et al., 2007), and Shinya Yamanaka's research group (Takahashi et al., 2007). Citing these successes, Professor Ian Wilmut of Edinburgh University, who is credited with producing Dolly the sheep by cloning, has decided not to pursue a license granted by British authorities to attempt to produce cloned human embryos for use in biomedical research. According to Wilmut, embryo-destructive means of producing the desired stem cells will not be necessary: "The odds are that by the time we make nuclear transfer [cloning] work in humans, direct reprogramming will work too. I am anticipating that before too long we will be able to use the Yamanaka approach to achieve the same, without making human embryos." Wilmut is quoted in Highfield, 2007. For a survey of possible non-embryo-destructive methods of obtaining pluripotent stem cells, see The President's Council on Bioethics, 2005.
2. E.g., in Boyle, 1979.
3. It is worth pointing out that contrary to a common misunderstanding, the Catholic Church does not try to draw *scientific* inferences about the humanity or distinctness of the human embryo from *theological* propositions about ensoulment. It works the other way around. The theological conclusion that an embryo is "ensouled" would have to be drawn on the basis of (among other things) scientific findings about the self-integration, distinctness, unity,

determinateness, etc., of the developing embryo. Contrary to another misunderstanding, the Catholic Church has not declared a teaching on the ensoulment of the early embryo. Still, the Church affirms the rational necessity of recognizing and respecting the dignity of the human being at all developmental stages, including the embryonic stage, and in all conditions. For a clear statement of Catholic teaching and its ground, see the Congregation for the Doctrine of the Faith, 1987: "[T]he conclusions of science regarding the human embryo provide a valuable indication for discerning by the use of reason a personal presence at the moment of this first appearance of a human life: how could a human individual not be a human person" (Section 5, I, 1, para. 3)?

4. My point here is not to make light of, much less to denigrate, the important witness of many religious traditions to the profound, inherent, and equal dignity of all members of the human family. Religious conviction can, and many traditions do, reinforce ethical propositions that can be rationally affirmed even apart from religious authority.
5. Thus, "recollecting (at her birth) his appreciation of Louise Brown [the first IVF baby] as one or two cells in his petri dish, [Robert] Edwards [said]: 'She was beautiful then and she is beautiful now,'" quoted in Finnis, 2000, 100. Edwards and his coauthor, Patrick Steptoe, accurately described the embryo as "a microscopic human being–one in its very earliest stages of development," (Edwards and Steptoe, 1981, 83). The human being in the embryonic stage of development is "passing through a critical period in its life of great exploration: it becomes magnificently organised, switching on its own biochemistry, increasing in size, and preparing itself quickly for implantation in the womb" (Ibid., 97).
6. Keith Moore and T.V.N. Persaud, in *The Developing Human: Clinically Oriented Embryology*, perhaps the most widely used embryology text, make the following unambiguous statement about the beginning of a new and distinct human individual: "Human development begins at fertilization when a male gamete or sperm (spermatozoon) unites with a female gamete or oocyte (ovum) to form a single cell–a zygote. This highly specialized, totipotent cell marked *the beginning of each of us as a unique individual*" (Moore and Persaud, 2008, 15, emphasis added).
7. A human embryo (like a human being in the fetal, infant, child, or adolescent stage) is not a "prehuman" organism with the mere potential to become a human being. No human embryology textbook known to me presents, accepts, or remotely contemplates such a view. Instead, leading embryology textbooks assert that a human embryo *is*–already and not merely potentially–a new individual member of the species *Homo sapiens*. His or her potential, assuming a sufficient measure of good health and a suitable environment, is to develop by an internally directed process of growth through the further stages of maturity on the continuum that is his or her life. Nor is there any such thing as a "pre-embryo." That concept was invented, as Lee Silver pointed out in his book *Remaking Eden*, for political, and not scientific, reasons (Silver, 1997, 39).

8. A cloned human embryo is not a subhuman organism. Cloning produces a human embryo by combining what is normally fused and activated in fertilization, that is, a properly epigenetically disposed human genome and the oocyte cytoplasm. Cloning, like fertilization, generates a new and complete, though immature, human organism. Cloned embryos therefore ought to be treated as having the same moral status, whatever that might be, as other human embryos. I respond to the arguments of my colleague on the President's Council on Bioethics, Paul McHugh, who claims that cloned embryos are not human beings but "clonotes," in the latter half of George and Lee, 2005.
9. The first one or two divisions, in the first thirty-six hours, occur largely under the direction of the messenger RNA acquired from the oocyte. Still, the embryo's genes are expressed as early as the two-celled stage and are required for subsequent development to occur normally. See O'Rahilly and Mueller, 2000, 38.
10. For a fuller explanation, see Lee and George, 2006.
11. For an entity to have a rational nature is for it to be a certain type of substance; *having a rational nature*, unlike, say, being tall, or Croatian, or gifted in mathematics, is not an accidental attribute. Each individual of the human species has a rational nature, even if disease or defect blocks its full development and expression in some individuals. If the disease or defect could somehow be corrected, it would perfect the individual as the kind of substance he is; it would not transform him into an entity of a different nature. Having a rational nature is, in Jeff McMahan's terms, a "status-conferring intrinsic property." So my argument is not that every member of the human species should be accorded full moral respect based on the fact that the more mature members have a status-conferring intrinsic property, as McMahan interprets the "nature-of-the-kind argument." See McMahan, 2005, 355 ff. Rather, my proposition is that having a rational nature is the basis for full moral worth, and *every* human individual possesses that status-conferring feature.
12. Clear-headed and unsentimental believers that full moral respect is due only to those human beings who possess immediately exercisable capacities for characteristically human mental functions do not hesitate to say that young infants do not deserve full moral respect. See, for example, Singer, 1995.
13. 'Not long ago, Peter Singer was asked whether there would be anything wrong with a society that bred children for spare parts on a massive scale. "No," was his reply (See Olasky, 2004).
14. Michael Gazzaniga has suggested that the embryo is to the human being what Home Depot is to a house, i.e., a collection of unintegrated components. According to Gazzaniga, "it is a truism that the blastocyst has the potential to be a human being. Yet at that stage of development it is simply a clump of cells.... An analogy might be what one sees when walking into a Home Depot. There are the parts and potential for at least 30 homes. But if there is a fire at Home Depot, the headline isn't 30 homes burn down. It's Home Depot burns down" (*Science* 2002, 1637). Gazzaniga gives away the game, however, in conceding, as he must, that the term "blastocyst" refers to a *stage of development* in the life

of a determinate, enduring, integrated, and, indeed, self-integrating entity. If we must draw an analogy to a Home Depot, then it is the gametes (or the materials used in cloning to generate an embryo), and not the embryo, that constitute the "parts and potential."

15. This conclusion would follow regardless of the acquired quality we chose as qualifying some human beings (or human beings at some developmental stages) for full respect.
16. For a more complete presentation of this argument, see Lee and George, 2005.
17. A further response to Bailey, and to similar arguments put forth by Lee Silver, may be found in George and Tollefsen, 2008, Chapter 6.
18. Recent research has raised questions about whether "brain death" is always equated with the irreversible loss of integral organic functioning; see Shewmon, 2001.
19. Lee and I have replied to other arguments that identify the human "person" as the brain or brain activity, and the human "being" as the bodily animal, in George and Lee, 2005.
20. Lee and I responded to Sandel in George and Lee, 2004.
21. Dr. William Hurlbut of Stanford University has pointed out that "[m]onozygotic twinning (a mere 0.4 percent of births) does not appear to be either an intrinsic drive or a random process within embryogenesis. Rather, it is a disruption of normal development by a mechanical or biochemical disturbance of fragile cell relationships that provokes a compensatory repair, but with the restitution of integrity within two distinct trajectories of embryological development." He goes on to explain that "the fact that these early cells retain the ability to form a second embryo is testimony to the resiliency of self-regulation and compensation within early life, not the lack of individuation of the first embryo from which the second can be considered to have 'budded' off. Evidence for this may be seen in the increased incidence of monozygotic twinning associated with IVF by Blastocyst Transfer. When IVF embryos are transferred to the uterus for implantation at the blastocyst stage, there is a two to ten-fold increase in the rate of monozygotic twinning, apparently due to disruption of normal organismal integrity" (Hurlbut, 2002).
22. For example, the plane of cleavage of the zygote predicts which cells will contribute to the inner cell mass and which will contribute to the trophectoderm (Plusa et al., 2005; Gardner and Davies, 2003; Rossant and Tam, 2004).
23. Lee and I presented this information in George and Lee, 2006.
24. This essay appeared in Daedalus 13 (2008), 23–35. It is reprinted with permission of MIT Press.

References

Bailey, R. 2001. Are stem cells babies? *Reasononline*. Available online at: www.reason.com/news/show/34840.html. Accessed 02 Sep 2010.

Boyle, J.M. 1979. That the fetus should be considered a legal person. *The American Journal of Jurisprudence* 59:59–71.

Congregation for the Doctrine of the Faith. 1987. Instruction on respect for human life in its origin and on the dignity of procreation: Replies to certain questions of the day [*Donum Vitae*]. http: www.vatican.va/roman_curia/congregations/cfaith/documents/rc_con_cfaith_doc_19870222_respect-for-human-life_en.html. Accessed 02 Sep 2010.

Denker, H.-W. 2004. Early human development: New data raise important embryological and ethical questions relevant for stem cell research. *Naturwissenschaften* 91:1–21.

Edwards, R., and P. Steptoe. 1981. *A matter of life*. London: Hutchinson's.

Finnis, J. 2000. Some fundamental evils in generating human embryos by cloning. In *Ethics and law in biomedical research*, ed. C.M. Mazzoni, 99–106. Dordrecht, The Netherlands: Kluwer.

Gardner, R.L., and T.J. Davies. 2003. The basis and significance of pre-patterning in mammals. *Philosophical Transactions of the Royal Society B: Biological Sciences* 358: 1331–1338.

George, R.P., and C. Tollefsen. 2008. *Embryo: A defense of human life*. New York, NY: Doubleday.

George, R.P., and P. Lee. 2004. Acorns and embryos. *The New Atlantis* 7:90–100.

George, R.P., and P. Lee. 2005. Dualistic delusions. *First Things* 150:5–7.

George, R.P., and P. Lee. 2006. The first fourteen days of life. *New Atlantis* 6:61–67.

Highfield, R. 2007. Dolly creator Ian Wilmut Shuns cloning. *Telegraph.co.uk*, November 16, 2007.

Hurlbut, W. 2002. Personal statement of Dr. William Hurlbut. In *Human Cloning and Human Dignity: An Ethical Inquiry.* Report of the President's Council on Bioethics, Washington, DC. Available online at: http://bioethics.georgetown.edu/pcbe/reports/cloningreport/appendix_html# hurlbut. Accessed 02 Sep 2010.

Lee, P., and R.P. George. 2005. The wrong of abortion. In *Contemporary Debates in Applied Ethics*, eds. A.I. Cohen and C. Wellman, 13–26. New York, NY: Blackwell Publishers.

Lee, P., and R.P. George. 2006. The first fourteen days of human life. *New Atlantis* 13:61–67.

McMahan, J. 2005. Our fellow creatures. *The Journal of Ethics* 9:353–380.

Moore, K., and T.V.N. Persaud. 2008. *The developing human: Clinically Oriented Embryology, 8th edition*. Philadelphia, PA: Saunders.

O'Rahilly, R., and F. Mueller. 2000. *Human embryology and teratology*. New York, NY: Wiley.

Olasky, M. 2004. Blue state philosopher. *World* 19. Available on-line at www.worldmag.com/articles/9987. Accessed 02 Sep 2010.

Piotrowska-Nitsche, K., Perea-Gomez, A., Haraguchi, S., and M. Zernicka-Goetz. 2005. Four-cell stage mouse blastomeres have different developmental properties. *Development* 132: 479–490.

Plusa, B., Hadjantonakis, A.K., Gray, D., Piotrowska-Nitsche, K., Jedrusik, A., Papaioannou, V.E., Glover, D.M., and M., Zernicka-Goetz. 2005. The first cleavage of the mouse zygote predicts the blastocyst axis. *Nature* 434:391–395.

President's Council on Bioethics. 2002. Session 5 meeting, January 18, 2002, Transcript available online at www.bioethics.gov/transcripts/jan02/jan18session5.html.

Rossant J., and P. Tam. 2004. Emerging asymmetry and embryonic patterning in early mouse development. *Developmental Cell* 7:155–164.

Sandel, M. 2004. Embryo ethics–the moral logic of stem cell research. *The New England Journal of Medicine* 351:207–209.

Science. 2002. *Metaphor of the Week* 295:1637.

Shewmon, D.A. 2001. The brain and somatic integration: insights into the standard biological rationale for equating 'Brain Death' with death. *Journal of Medicine and Philosophy* 26: 457–478.

Silver, L. 1997. *Remaking eden*. New York, NY: Avon Books.

Singer, P. 1995. Killing babies is not always wrong. *The Spectator* 16:20–22.

Stanton, J.A., Macgregor, A.B., and D.P. Green. 2003. Gene expression in the mouse preimplantation embryo. *Reproduction* 125:457–468.

Takahashi, K., Tanabe, K., Ohnuki, M., Narita, M., Ichisaka, T., Tomoda, K., and S. Yamanaka. 2007. Induction of pluripotent stem cells from adult fibroblasts by defined factors. *Cell* 131:1–12.

The President's Council on Bioethics. 2005. White paper: Alternative sources of pluripotent stem cells. http://bioethics.georgetown.edu/pcbe/reports/white_paper/text.html. Accessed 07 Sep 2010.

Torres-Padilla, M., Parfitt, D., Kouzarides, T., and M. Zernicka-Goetz. 2007. Histone arginine methylation regulates pluripotency in the early mouse embryo. *Nature* 445:214–218.

Vogel, G. 2005. Embryologists polarized over early cell fate determination. *Science* 308: 782–783.

Yu, J., Vodyanik, M.A., Smuga-Otto, K., Antosiewicz-Bourget, J., Frane, J.L., Tian, S., Nie, J., Jonsdottir, G.A., Ruotti, V., Stewart, R., Slukvin, I.I., and J.A. Thomson. 2007. Induced pluripotent stem cell lines derived from human somatic cells. *Science* 318:1917–1920.

Chapter 4
Compassion and the Personalism of American Jurisprudence: Bioethical Entailments

R. Mary Hayden Lemmons

Prima facie, compassion seems to dictate the permissibility of assisted suicide for the terminally ill as well as the permissibility of in vitro fertilization for desperate would-be parents. But what are the grounds for identifying these acts as compassionate? Could it not be the case that compassion proscribes assisted suicide as well as the use of in vitro fertilization? This article considers this question by examining the nature of compassion as well as Joseph Boyle's and Germain Grisez's personalist arguments from American jurisprudence.

Compassion is that state in which one feels somehow united to one who is suffering and wills the alleviation of that suffering. As such is it both an emotional state as well as an intentional state. Mothers who wrap their sympathetic arms around their crying toddler with the skinned knee carry the child to the band-aids and antibiotic. Fathers who urge their disheartened baseball players not to let a losing score determine the next inning not only sympathize with their players but take action to move them beyond their pain. Nurses who urge their patients to walk after a painful surgery are showing compassion as well; paradoxically, by encouraging their patients to endure the increased pain necessary for regaining health.

So does compassion require family members and medical professionals to assist in the suicide of terminal patients or in the alleviation of the pain of childlessness through in vitro fertilization? Or is it the case that compassion has certain necessary conditions without which the otherwise compassionate act is actually an act of hatred? Consider the case of a grandson poisoning his wealthy grandfather so that he can inherit earlier rather than later. Even if the grandfather were terminally ill and in terrible pain, the act would not be one of compassion but of selfish gain because the intention was not to benefit grandfather by alleviating pain but to inherit quickly. The intention of benefiting the sufferer, or the patient, is thus a necessary condition of compassion.

R.M. Hayden Lemmons (✉)
University of St. Thomas, St. Paul, MN, USA
e-mail: rmlemmons@stthomas.edu

C. Tollefsen (ed.), *Bioethics with Liberty and Justice*, Philosophy and Medicine 110,
DOI 10.1007/978-90-481-9791-0_4, © Springer Science+Business Media B.V. 2011

4.1 In Vitro Fertilization

Is the condition of patient benefit met in those cases where compassion seems to dictate the permissibility of in vitro fertilization because a couple desperately wants a biological child? On the one hand, it might seem that the condition is met because the adults get what they want: namely, an embryonic child to love and to rear until adulthood. However, just as in cases of organ transplant where there is more than one patient (the donor and the donor recipient), in cases of in vitro fertilization there is also more than one patient (the egg donor, the sperm donor, the would-be mother, and the embryo).

Is it, however, begging the question to identify an embryo as a patient? If it were, it would be inconceivable that in the future, genetic manipulation of embryonic DNA could ever result in curing post-puberty genetic diseases or in providing post-puberty genetic "enhancements." Likewise it would be inconceivable for a future doctor to be sued by a young adult for so botching his embryonic genetic manipulation as to cause a genetic deformity to emerge during puberty. Clearly then embryos and adults are human beings in different stages of development as shown by the control asserted over their self-development by their own DNA.[1] In vitro fertilization accordingly subjects both adult and embryonic human being to procedures that makes them patients. Accordingly, whether in vitro fertilization is a compassionate procedure can be ascertained by determining whether it benefits all of its patients. We shall, however, concede for the sake of argument the rather doubtful claim that all the adults are benefitted and focus solely on the question of the benefit to embryos.

Not every embryo is benefitted by being implanted, born, and reared by a loving couple; for, this procedure does not typically implant every embryo. "Spare" embryos are given, by their parents, to researchers to be experimented upon; others are put into the legal limbo of being frozen for possible later implantation; others are killed before implantation; and others are implanted only to be selectively aborted as unwanted. Being conceived through in vitro fertilization thus dramatically increases the likelihood of being harmed—or, even killed—before being born. For these reasons, in vitro fertilization is not an act of compassion for most of the embryos involved.

Neither is it a process that is fair, since it typically involves permitting only one or two embryonic siblings to be carried to term and born, while the other siblings are killed, frozen or turned over for experimentation. In brief: in vitro fertilization proceeds by giving some embryos a normal life at the cost of depriving others of a normal life. To so discriminate is unfair to those deprived of the chance to develop into adults. In vitro fertilization thus mistreats some human beings for the sake of others. This means that the philosophical warrant for in vitro fertilization requires an instrumentalist bioethics that treats some human beings as instruments for the satisfaction of the desires of other human beings.

But does in vitro fertilization mistreat human *persons*—is the embryonic human being also a person? Any attempt to define human beings in ways that make it possible to be a human being but not a person involves defining a substance through activities that may or may not be present. For instance, Michael Tooley defines a person in terms of self-consciousness (Tooley, 1972).[2] Such definitions run aground when living persons lack the so-called defining activity or property. For instance, we all lack self-consciousness when we are asleep, but lose neither our personhood nor our rights while we sleep. Tooley's definition, moreover, cannot be saved by claiming that we are persons only because we are capable of waking up and becoming self-conscious; because, then it is not the property of self-consciousness that defines us but rather the substance that makes that property possible. What is this substance? Simply being a living human being: otherwise it would not be possible for humans suffering from hypothermia, or from barbiturate overdose, to recover after satisfying the 1968 Harvard Criteria of Brain Death.[3] In other words, since the properties of those who are alive but suffering from hypothermia or barbiturate overdose are indistinguishable from the properties of the brain dead, none of these properties suffice to identify living persons. This means that it is an error to identify "person" with some supervening property of being human rather than with the very essence of being human, i.e, with being a human substance. As explained by Boyle:

> [P]erson in its ordinary use ... falls into the category of substance—telling us what something is. A person is an individual of a certain kind. It refers to individuals insofar as these individuals are members of a certain natural kinds or species. It does not refer to them insofar as they have some degree of certain properties in addition to those essential features which define them as members of the species in question. This feature of the semantics of 'person' is important for the functioning of the notion of 'legal person,' especially insofar as this notion is used to determine which things have rights. Presumably, one either has rights or one does not; one either is or is not a legal person. Likewise one is either a member of a species or one is not (Boyle, 1979, 65–66).

The personhood of the developing human embryo whereby it is a human being suffices to identify the failure to ban in vitro fertilization as a failure to extend equal protection of the law to human persons. This means that the current practice of in vitro fertilization entails an instrumentalist jurisprudence that identifies a practice of instrumentalist bioethics as legally permissible.

Instrumentalist bioethics threatens to transform medicine from a profession with internal standards based on health to those of a business ruled by the desires of clients. Consider, for example, that in vitro fertilization requires medical professionals to be willing to manufacture embryos, implant some, freeze some, kill some, and even reserve some for later experimentation. In this nexus of values, there is the reduction of medical professionals to well-paid technicians doing the will of their clients without regard for the health and well-being of all involved: not only are embryos being treated as instruments of the client's desires but so are the medical professionals. To reduce medical professionals to the status of tools for paying clients is especially pernicious because it ends medicine's status as a profession defined by the requirements of health and identifies hospitals and clinics as mere instruments of social utility.

Thus, as the grip of instrumentalist bioethics tightens, we can expect the law to increasingly succumb to instrumentalist conceptions and become ever more intolerant of medical professionals whose moral beliefs are contrary to that of their clients.

For instance, in 2005, lawmakers in the state of Washington introduced a bill to compel pharmacies and pharmacists to dispense whatever is prescribed.[4] The attitude that health care providers are at the service of whatever a client wishes has increasingly been extended to Catholic hospitals who see health care as a Christian ministry, rather than as a way to please clients, maximize profit or enable workers to return to full productivity. For instance, on January 18, 2008 Fox News reported that a Catholic hospital was being sued by a transgender woman for refusing to do breast enlargement surgery (Fox News.com, 2008).

To treat human beings and their institutions as mere instruments of profit, utility, or desire is incompatible with the traditional view of human beings as having a nature with an intrinsic dignity that transcends their value to others. If being a human embryo is insufficient for having a transcendent dignity, then no one has any intrinsic and transcendent dignity simply as a human being. If no one has any intrinsic and transcendent dignity, then the value of human life is conditioned on something extrinsic, e.g., being loved by another or being useful. If the value of human life is conditioned on something extrinsic, then the treatment of any human being is always a function of a cost-benefit analysis where one's ability to satisfy the desires of others determines one's treatment—and one's value. This is the world of instrumentalist bioethics where the cost of every treatment is weighed either against the usefulness of the patient for society or towards thickening the practitioner's wallet. Surely such considerations constituted the mindset of those involved in stealing kidneys in India[5] and of those who amputated healthy limbs upon the request of their patients.[6]

Although instrumentalist bioethics is making inroads in the United States, it cannot succeed without overturning American jurisprudence. This is because, as pointed out by Joseph Boyle and Germain Grisez, the foundations of American jurisprudence are ultimately based on the consent of those who have joined together as a people to establish a political structure that enables them to discharge their moral responsibilities to others—including the weak (Grisez and Boyle, 1979, 29). This argument from responsible citizenship can be summarized by the following claims:

First, the preamble of the U.S. Constitution shows that our government is based on the free consent of those who are neither naturally a slave nor naturally a ruler but equals.[7]

Second, the free consent of the governed must be a moral consent for otherwise the governed would not be truly governed by law but by fear of punishment or by a calculation of self-interest (Grisez and Boyle, 1979, 30).[8] (The law's authority must be moral authority in order to deserve obedience when punishment is not likely and self-interest is not at stake.)

Third, moral consent to the Constitution would not be moral if it were not an exercise in moral responsibility: "Because my responsibility for my own life and other goods, and for the security of others for whom I care, demands that I obtain protection" (Grisez and Boyle, 1979, 29). Among those for whom I am morally responsible are those who are weak that I can help (Grisez and Boyle, 1979, 35). For instance, adults are morally obligated to rescue drowning toddlers in wading pools; teachers are morally obligated to prevent stronger students from bullying the weaker ones and so forth.

Fourth, as an exercise in moral responsibility, the Constitution established on the basis of moral consent must not only protect the goods that brings one into the political union,[9] but must also protect the persons for whom one is responsible—including the weak.[10] Providing such protection was a key reason for the Constitution's ninth and tenth amendments.[11] It was also the reason why the fourteenth amendment to the Constitution was ratified in 1868, namely, to protect ex-slaves from the unfair laws being passed by post-Civil War southern states.[12] This emphasis on protecting persons identifies American jurisprudence as personalistic, i.e., as valuing persons for their own sake.

Fifth, "the only thing common to all already recognized by law as natural persons is membership in the human species" (Grisez and Boyle, 1979, 304). Indeed, as argued by Boyle, the use of criteria for personhood that is narrower than membership in the species is necessarily sectarian (Boyle, 1979, 69–70); it is to adopt secular humanism. In his book with Grisez, Boyle notes that secular humanism was defined as a religion by the U.S. Supreme Court in *Torcaso v Watkins* (Grisez and Boyle, 1979, 301).

Sixth, to exclude any member of the human species whether embryonic, terminally ill or less than perfect from the equal protection of the law is to attack the equality that underpins the Constitution. It is to practice a particularly vicious form of discrimination, since it involves the denial of a person's rights. As argued by Boyle, to adopt any criterion of human personhood other than conception is to practice arbitrary discrimination, often for self-serving purposes (Boyle, 1979, 66). There is no warrant for this kind of arbitrariness, since it "is an arbitrariness that deprives people of basic rights—one that no one would endorse supposing that he or something which developed into him might be one of the individuals just beyond the precise point which was chosen to be the cutoff line for having the relevant capacity" (Boyle, 1979, 68). In brief, as put by Boyle and Grisez: "[D]iscrimination which rationalizes killing is equally vicious whether it is rooted in an ideology of racial perfection or in an ideology of individualistic perfection, which asserts that no unwanted child should ever be born and no life below a certain standard of quality should any longer be protected" (Grisez and Boyle, 1979, 304).

Seventh, instrumentalist bioethics and jurisprudence necessarily requires an arbitrary and vicious form of discrimination that excludes giving every member of the human species equal legal protection from being killed.

Eighth, therefore, instrumentalist bioethics and jurisprudence are contrary to the U.S. Constitution.

The implication of this argument from responsible citizenship is not that it is some narrow sectarian viewpoint that rejects instrumentalist bioethics and jurisprudence but that it is rather the very *raison d'etre* of the American Constitution. From this, it follows that the current laws that permit the killing of human beings whether embryonic as in in vitro fertilization, fetal as in abortion, or terminally ill as in Oregon's law of assisted suicide have placed American jurisprudence at odds with itself.

Against this argument from responsible citizenship is the objection that since the goods sought by America's founders include freedom and the right to pursue happiness, Americans have the right to commit suicide or to implant embryos, and to have medical professionals help them do so—even if against their will. Instrumentalist bioethics and jurisprudence thus breeds the insolent conviction that it is not unjust to treat medical professionals as tools without consciences and the right to honor their consciences. Underpinning these assumptions are several others, namely, that suicide is a means to happiness, that killing oneself is an act of empowerment rather than an act of self-annihilation, and that neither rights nor happiness have objective criteria (they are whatever one decides them to be). Are these assumptions true?

4.2 The Bioethics of Euthanasia

Suicide is typically committed by those who see it as saving themselves from suffering—not as intensifying their suffering.[13] To see suicide as a means of release requires separating one's identity from one's body: the body is causing me intolerable pain so I must separate it from myself. The body is not me but rather my instrument to be disposed of whenever I choose. From this perspective, death frees one of an instrument of torture and is a blessing. So is it the case that assisted suicide is an act of compassion for the terminally ill? Does death benefit the terminally ill? Instinctively, most of us would answer affirmatively because we see death as ending suffering. But is this not to mistake a coincidence for a cause? After all, if a person is not suffering, death is not seen as a blessing but as an evil. Accordingly, it is not possible to view death as a blessing without also adopting the view that it is possible for a life not to be worth living. Otherwise, the compassionate heart would not seek to kill the suffering but to provide adequate pain relief and whatever kind of support was needed, whether financial, emotional or spiritual.

Two questions now arise: (1) Is it possible for a life not to be worth living? (2) Is it possible for a person to have no value? Proponents of euthanasia argue that a life of misery or incurable suffering is not worth living—as if that verdict does not also identify a person as being worthless. Euthanasia proponents can do this because they adopt an instrumentalist bioethics that includes the instrumentalist view of the body wherein a body lacks personal identity and may be treated however its owner wishes. In these assumptions, we see not only the establishment of a certain dualism that divorces the body from the person but also a devaluation of the body such that bodily life has no value apart from what is bestowed upon it by personal choices. Entailed by this position is the notion that all value is bestowed by beings capable of personal choices: there are no necessary or intrinsic values. From this perspective, human beings unable to choose due to illness or being too young to choose have no intrinsic value—no intrinsic right to live. They are to live or die at the pleasure of others.[14] Also from this perspective, bodies that are in pain or that are incurable are not worth preserving—even if they are human bodies; human suffering can have no personal value and thus should be legally terminated whenever costs are greater than benefits. This reliance on cost-benefit analysis as the sole moral and legal determinant identifies utilitarianism as the morality of instrumentalist bioethics and a legal proceduralism that identifies the law with whatever promotes the will of citizens while according with democratic procedures lacking in substantive values.[15]

The moral and juridical assumptions made by instrumentalist bioethics can be successfully challenged in two ways. On the one hand, all the resources of the philosophical canon can be brought to bear on these assumptions. For instance, the instrumentalist dualism of instrumentalist bioethics could be challenged by the hylomorphic tradition of Aristotle and Aquinas; its utilitarianism could be challenged by the classical natural law tradition originating in ancient Greece and Rome; and, its instrumentalist jurisprudence warranted by an empty formalistic proceduralism could be challenged by Boyle's argument from responsible citizenship. The problem with this canonical approach is that it is beyond the abilities of most Americans who must adjudicate these issues for themselves and decide whether, for instance, to use in vitro fertilization, to abort, to commit suicide, to support euthanasia, or to fund the manufacture of human embryos for research or implantation or freezing as spare children for the future. Fortunately, Boyle offers a more accessible alternative by focusing on common human experiences. This focus suffices to characterize his argument as personalist.[16] It also suffices to prove that everyone has the experiences necessary to prove what I'm calling instrumentalist bioethics false.

4.3 Personalism and the Falsity of Instrumentalist Bioethics

Instrumentalist bioethics assigns only an instrumental value to the body.[17] And although compatible with materialism and the claim that persons have only an instrumental value, instrumental bioethics is more likely to be placed within the context of dualism, where the body is treated merely as an instrument of the person and not as a part of the self. Opposing both materialistic and dualistic instrumentalism is Boyle's personalism, which claims that the value of the person transcends all instrumentality and that human persons are also bodies. This type of personalism looks within human experiences and notes that we experience our bodies as part of ourselves. We do not experience our bodies as mere instruments of self-gratification but as ourselves. For instance, we say: "I *am* sleepy and don't hit *me*." Or, as argued by Boyle and Grisez, we experience ourselves as "unified wholes:"

> This awareness is especially clear in carrying on purposeful human acts. Lawyers examine cases, engaging hands and eyes, memory and thought ... Lawyers experience all of these

activities as their own; they are directly aware of what they are doing as integrated action, as part of their single and only life (Grisez and Boyle, 1979, 378).

In other words, if dualism or instrumentalism were true and if personalism were not true, no one could experience both sensations and personal choices as one's own.

Common human experience also disproves dualism in a second way. We can be embarrassed by our bodies: acne, for instance, makes teenagers cringe as does the loss of bladder control in the sick. But, as Boyle and Grisez point out, "Consistent dualists should say that their own living bodies simply have nothing to do with personal dignity" (Grisez and Boyle, 1979, 377).

A third argument against dualism takes seriously the divorce of the body from the person and notes that it denies that bodily life is constitutive of personal acts—as if personal acts were not acts of living human beings. Yet singing, for instance, is a personal act inseparable from the body and its life: bodily life is a necessary condition for singing. As pointed out by Boyle and Grisez:

> Life is not merely one process among others, which might be distinguished from breathing, feeling pain, choosing, talking, and administering treatment. The life of a living entity is indistinguishable from the very reality of the entity.... Thus human activities ... are not separate from a person's life. Life is not a characteristic of one part of a whole and these activities properties of some other part of it (Grisez and Boyle, 1979, 377).

"Thus," according to Boyle and Grisez, "the underlying, pervasive, and inclusive character of life with respect to all of a person's activities is a fact, and this fact is inconsistent with any form of dualism" (Grisez and Boyle, 1979, 278). In other words, if personal acts are acts of the living body and acts of the living body are personal acts, then not only is dualism false but so is the bioethics that presupposes dualism.

These three personalist arguments on the basis of common experience excel in enabling the typical citizen not only to ascertain the falsity of dualism and the instrumentalist view of the living human body, but also to ascertain the falsity of the instrumentalist bioethics that presupposes such dualism. In brief: if dualism with its instrumentalist view of the body were true, then we would not be able to experience our bodily actions as a part of us, we would never be embarrassed by our bodies, and we would not be able to experience our living acts as personal. These experiences disprove dualism and prove the necessity of a hylomorphic personalism that considers the person to be a body–albeit not only a body.

But, perhaps, one may object that the falsity of dualism only disproves an instrumentalist view of the body but not the instrumentalist view of the person. Two responses.

One the one hand, the objection is technically correct in claiming that disproving dualism does not disprove the instrumentalist view of the person. The falsity of dualism proves only that the body and bodily life has the same value as the person. So it remains logically possible that the person has only an instrumental value and lacks unconditional value. Arguments against assigning only instrumental value to the person can be mounted by Kantians (since persons are free to choose their own ends, persons have unconditional worth); by hylomorphists like Aristotle or Aquinas (since human persons have souls that transcend the limitations of matter as shown by the ability to know universals and by free choice, persons have unconditional worth); by theists (since human persons image God—or otherwise partake in the divine, human persons have unconditional value); and, by Christians (since Christ died for human beings, persons have unconditional worth). In this regard, it is interesting to note that Kant's argument has been turned against him by those who argue that only those human beings capable of free choice have unconditional value, thereby excluding infants, the senile, and the comatose. This twisting of Kant's argument is only possible through ignoring what has been called Kant's principle of humanity: humanity within oneself or another is always to be respected and never reduced to a mere means.[18] In any case, although I think that each of these arguments is successful when carefully unpacked, I am unsure of their persuasiveness in today's public arena.

On the other hand, the objection that human persons lack unconditional value can be accepted only by rejecting, or ignoring, the personalist facts used by Boyle and Grisez in establishing the falsity of dualism. These facts reveal that we experience the world from a personalist perspective, that is, we experience the world from our own personal perspective. You can accordingly, as the popular song *Unwritten* by Natasha Bedingfield puts it, "Release your inhibitions. Feel the rain on your skin. No one else can feel it for you." Only persons experience the world in this way. Only persons can release their inhibitions, freely choose to feel the rain, and relish the experience. In other words, the decision to go play in the rain is a personal act to experience rain for one's own sake; it is a free choice that affirms that the self is one's own end, even while also acknowledging that self-transcendence is necessary for self-fulfillment.[19] Nevertheless, free choice reveals that the self's value is not extrinsic, but intrinsic and unconditional. No person can thus be treated as having only instrumental value or as lacking in an unconditional value without being mistreated.

Free choices, moreover, not only structure one's life but make one a certain type of person. Free choice is the power of self-determination. As put by Boyle:

> Free choice is understood by those who affirm it as that act by which the person most fully self-determines himself. It is in making free choices that a person determines himself to be the sort of person he or she is. If choice is free, then the person in choosing is not determined by something other than himself—or even by determining features within his own nature—to be a certain kind of person; rather, the person in choosing freely determines himself or herself (Boyle, 1980, 189).

Since free choices shape one's life, they determine whether one is kind or nasty, courageous or fearful, generous or stingy. It is one's choices that determine one's attitude towards what cannot be changed—especially when suffering and death is imminent. Since this is so, unchosen suffering cannot determine one's meaning—or one's value: it is not suffering and dying that determines one's identity but how one chooses to react to suffering and dying. Boyle's personalism thus argues that suffering cannot annul the value of one's life and that compassion requires helping even the dying retain the awareness and appreciation for the value of life. But is personalism really true?

Without doubt, many experience their suffering as negating their value and rendering their life meaningless. Are they victims of a faulty mind-set? If so, it would not be compassionate to reinforce their sense of worthlessness by providing them with the instruments of suicide and by creating an environment where the dominant beliefs are that the suffering are better off dead and that involuntary euthanasia for the comatose is a kindness.

To determine the truth about a life of suffering one must consider that if it were true that suffering negates personal dignity, then it would not be possible for anyone suffering in a hopeless situation to be dignified—or, to affirm a transcendent personal dignity. Yet Viktor Frankl testified that even in the most dismal Nazi concentration camp, it was possible for one to find meaning and dignity not only while suffering, but also through suffering.[20] Consequently, camp inmates encouraged each other not to commit suicide (Frankl, 1959, 100), but to commit to the irreplaceable value of each individual life—despite any feelings of worthlessness and hopelessness. How were they able to avoid suicidal desires? According to Frankl, they did it either by identifying a future goal worth the struggle against their suffering or by identifying their suffering as meaningful insofar as it gave them opportunities to reach out to others or to resist descending to the status of animals ruled by pleasure and pain.[21] Identifying their current lives of suffering as meaningful sufficed to avoid despair and suicide. The key to achieving a meaningful life was realizing that one's choices make one into a certain type of person—even when suffering and dying, or dealing with those who are dying. Every moment is a test: what are my values? Shall I affirm or negate my life?

The experience of Viktor Frankl teaches three lessons. First, it is not the case that suffering necessarily negates human dignity and the value of one's life. It is possible to lead a meaningful life when suffering even in a concentration camp—or a contemporary hospital or hospice. Second, it is a great kindness to help others find a life-sustaining meaning to their suffering. Three, it is not compassionate to reinforce a sufferer's experience of meaninglessness by facilitating his or her suicide or by surrounding him or her with those who believe that he or she is without value—without dignity now that suffering and death beckons.

Moreover, it should be noted that if it were possible for any human life to be without value—without dignity—then it would not be possible for any human life to have intrinsic dignity. If none have intrinsic dignity, then none would ever have that experience—which is so common—of seeing the beloved and exclaiming "truly you are more wonderful than all the world's beauty and sweetness." Nor would anyone ever have the urge to risk death fighting to defend their families or their country or their faith. Nor would anyone ever be moved to compassion by the misfortune of others. In other words, the affirmations of love would not even be conceivable if nothing had intrinsic dignity. Indeed, both benevolence and beneficence arise from a love that seeks to affirm that the other is a person worthy of being loved for his own sake. Love even makes it possible to find happiness amidst suffering (Lemmons, 2010).

A final argument against the instrumentalist perspective that denies that embodied human persons have intrinsic dignity can be mounted on the basis of human interdependencies whereby a person can neither be born nor live without benefiting from the actions of others and whereby cooperation leads to a better life for all. Cooperation requires mutual commitment to a common goal, e.g., building this barn, this freeway, this nation. This mutual commitment establishes an equality—and, in the United States, a friendship (Pakaluk, 1994)—between those cooperating that is incompatible with the inequality caused by one person treating another as a mere instrument.[22]

The equality caused by the commitment of the American citizen to establish, as put by Abraham Lincoln,[23] a "government of the people, by the people, and for the people" cannot abide the inequality so necessary to the instrumentalist view of the person, instrumentalist bioethics and instrumentalist jurisprudence.[24] For all these forms of instrumentalism deny the intrinsic dignity and the rights of some persons for the benefit of others—be they researchers in need of embryos; parents in search of biological children now and possibly in the future; family members either worn out by care-giving or scared of care-giving commitments; or, health care providers seeking to maximize profitability. To deny the intrinsic dignity of any human being is to deny the human dignity of all. Instrumentalist bioethics and jurisprudence are thus not only insulting to the suffering and dying, but to us all.

It is therefore time to end the Oregon experiment with assisted suicide as an affront to American jurisprudence as well as to ban embryo farming and *in vitro fertilization*. After all, *death with dignity* should have only one meaning in America, namely, that the dying receive adequate pain relief,[25] the physical, emotional and spiritual support necessary, and the comfort of knowing not only that they were too loved to be killed like a dried up milking cow, but also that none of their embryonic children were so despised as to be killed, frozen, conceived, or experimented on for the sake of others. True compassion demands no less.

Notes

1. That adults and embryos are the same being at different stages of development was argued in George (2002, 317–324).
2. "Let us turn now to the first and most fundamental question: What properties must something have in order to be a person, i.e., to have a serious right to life? The claim I wish to defend is this: An organism possesses a serious right to life only if it possesses the concept of a self as a continuing subject of experiences and other mental states, and believes that it is itself such a continuing entity . . . this claim, . . . I will call the self-consciousness requirement" (Tooley, 1972, Reprinted 1999, at 24).
3. The Harvard Criteria lists four criteria of brain death, namely, (1) unreceptivity and unresponsivity to stimuli; (2) no spontaneous movements or spontaneous breathing; (3) no reflexes; and (4) flat electroencephalogram. It then states: "The validity of such data as indications of irreversible cerebral damage depends on the exclusion of two conditions: hypothermia (temperature below 90 degrees

Fahrenheit) . . . or central nervous systems depressants, such as barbiturates" Committee of the Harvard Medical School to Examine the Definition of Brain Death (1968, Reprinted 1999, at 288).

4. See Zwillich (2005). Two Washington pharmacists and a pharmacy owner subsequently filed suit in a federal court saying that the new law requiring them to sell contraceptives violates their civil rights by forcing them into choosing between "their livelihoods and their deeply held religious and moral beliefs" (Woodward, 2007).
5. On February 1, 2008, *Newsweek* reported that, in India, police have busted an organ-theft ring that "apparently involved four doctors, five nurses, 20 paramedics, three private hospitals, 10 pathology clinics and five diagnostic centers, police believe, and drew patients from as far afield as Canada, Greece and the United States" (Overdorf, 2008).
6. For a discussion from the *New York Times* see Henig (2005). For a 2006 news report from the BBC that three doctors in India offered to amputate the healthy limbs of beggars for about $200; see Sudworth (2006).
7. "[T]he Preamble to the Constitution of the United States makes fully explicit the dependence . . . upon consent . . . 'We the People of the United States . . . do ordain and establish this Constitution'" (Grisez and Boyle, 1979, 27). Indeed, the formation of political society is an exercise of liberty (45). The object of this consent is the "justice of the constitutional framework" (39).
8. Grisez and Boyle (1979, 47) give further support to the claim that consent of the governed must be a moral consent by analyzing liberty. "To deny that liberty is a good which makes a moral demand is to admit that slavery might be morally acceptable. . . .[I]t is an injustice for the state to infringe upon legitimate liberties."
9. "Governments are instituted to protect the unalienable rights to life, liberty and the pursuit of happiness" (Grisez and Boyle, 1979, 34); "The public interest extends only to those goods the cooperative pursuit of which gives political society the unity it has" (38). It is possible to argue that since the United States was established to protect an innocent person's life, liberty and the pursuit of happiness, the destruction of that life through euthanasia cannot be legalized without abrogating not only the very reason for the state's existence but also its moral authority. In tacit recognition of this point, instrumentalist bioethicists deny that a living human body suffices to establish personhood and the rights of personhood. Boyle (1979, 60) argues to the contrary: "[T]he proposition that I wish to defend . . . is that the fetus should be treated as a 'person' in the sense of that phrase that is relevant for legal discussion."
10. "Indeed, the very concern for political equality which entails the demand that government be by consent of the governed also tends to generate a demand for the extension of government into those areas . . . where exploitation of the weak by the strong is likely to occur" (Grisez and Boyle, 1979, 34).
11. Consider, for instance, these two amendments: Amendment IX: "The enumeration of the Constitution of certain rights shall not be construed to deny or disparage others retained by the people." Amendment X: "The powers not

delegated to the United States by the Constitution, nor prohibited by it to the States, are reserved to the States respectively, or to the people."

12. The first section of the fourteenth amendment reads: "All persons born or naturalized in the United States and subject to the jurisdiction thereof, are citizens of the United States and of the State wherein they reside. No State shall make or enforce any law which shall abridge the privileges or immunities of citizens of the United States; nor shall any State deprive any person of life, liberty, or property, without due process of law; nor deny to nay person within its jurisdiction the equal protection of the laws."
13. See, for instance, Lavery et al. (2001). This article argued that the desire for euthanasia or assisted suicide "resulted from fear and experience of two main factors: disintegration and loss of community."
14. Peter Singer (1993) argues that being human is irrelevant to the morality of killing: "In Chapter 4 we saw that the fact that a being is a human being, in the sense of a member of the species Homo sapiens, is not relevant to the wrongness of killing it; it is, rather, characteristics like rationality, autonomy, and self-consciousness that make a difference. Infants lack these characteristics. Killing them, therefore, cannot be equated with killing normal human beings, or any other self-conscious beings. This conclusion *is not limited* to infants who, because of irreversible intellectual disabilities, will never be rational, self-conscious beings.... [S]uppose that a newborn baby is diagnosed as a haemophiliac. The parents, daunted by the prospect of bringing up a child with this condition, are not anxious for him to live. Could euthanasia be defended here?...When the death of a disabled infant will lead to the birth of another infant with better prospects of a happy life, the total amount of happiness will be greater if the disabled infant is killed. The loss of happy life for the first infant is outweighed by the gain of a happier life for the second. Therefore, if killing the haemophiliac infant has no adverse effect on others, it would, according to the total view, *be right to kill him*" (186, emphasis mine).
15. It is a mistake to view democratic procedure as a mere process for expressing the will of the majority, i.e., as lacking in substantive values. Perhaps the best way to see this is to consider that it is possible for the will of a people to reject democratic processes as too unwieldy and inefficient. Such a divorce between democratic processes and the will of the people indicate that the ground of democratic processes is not simply as a convenient and arbitrary convention or tool for ascertaining a people's will but actually a substantive commitment to democratic values that require the free participation of citizens as well as their commitment to the rule of law, the equal protection of the law and responsible citizenship. For instance see Fuller (1969, 182–186), wherein he argues that the procedures of law presuppose substantive moral commitments that include the love of neighbor.
16. Personalism designates an approach to philosophy that emphasizes the irreducible value of the person rather than formulating a complete philosophical system. As a result, there are different types of personalisms depending upon the systematic stances adopted. The type of personalism characteristic of Boyle

is a hylomorphic personalism in the tradition of Aristotle and Aquinas that was called Thomistic Personalism by Karol Wojtyla in his 1961 paper "Personalizm tomistyczny."

17. Instrumental value is typically assigned to the body by identifying the body as personal property that can be done with as one wills. Judith Jarvis Thomson (1977), for instance, bases her pro-abortion arguments on the presupposition that since one has a prior claim to one's own body, "the mother has a right to decide what happens in and to her body" (48).
18. "Act in such a way that you treat humanity, whether in your own person or in the person of another, always at the same time as an end and never simply as a means" (Kant, 1993, 36).
19. One acknowledges that one is one's own end by seeking the fulfillment of transcending the self through knowing and loving neighbors—and, most notably, God. For this reason, Aquinas argues that self-perfection is the ultimate end of all things while also arguing that there is no ultimate self-perfection without knowing and loving God. See St. Thomas Aquinas (1947), I–II, Q. 1, a. 7: "all desire the fulfillment of their perfection." Also see I–II, Q. 1, a. 8: "the end is twofold, . . . the thing itself in which is found the . . . good and the use or acquisition of that thing. . . . If, therefore, we speak of man's last end as of the thing which is the end, . . . God is the last end. If, however, we speak of man's last end, as the acquisition of the end, then . . . [it is] knowing and loving God." I have published on this point under my maiden name; see, for example, Hayden (1989, 1991).
20. "Life in a concentration camp tore open the human soul and exposed its depths" (Frankl, 1959, 108); "Of the prisoners only a few kept their full inner liberty and obtained those values which their suffering afforded, but even one such example is sufficient proof that man's inner strength may raise him above his outward fate. Such men are not only in the concentration camps. Everywhere man is confronted with fate, with the chance of achieving something through his own suffering" (88–89).
21. "Woe to him who saw no more sense in his life, no aim, no purpose, and therefore no point to carrying on.. . . The typical reply with which such a man rejected all encouraging arguments was, "I have nothing to expect from life anymore." What sort of answer can one give to that? What was really needed was a fundamental change in our attitude toward life. We had to learn ourselves and furthermore, we had to teach despairing men, that it did not really matter what we expected from life, but rather what life expected from us" (Frankl, 1959, 98); "When a man finds that it is his destiny to suffer, he will have to accept his suffering as his task.. . . His unique opportunity lies in the way in which he bears his burden. For us, as prisoners, these thoughts . . . kept us from despair, even when there seemed to be no chance of coming out of it alive" (99).
22. "When two different people consciously choose a common aim this puts them on a footing of equality, and precludes the possibility that one of them might be subordinated to the other. Both . . . are as it were in the same measure and to

the same extent subordinated to that good which constitutes their common end" (Wojtyla, 1993, 28); "[E]very person is by nature capable of determining his or her aims. Anyone who treats a person as the means to an end does violence to the very essence of the other, to what constitutes its natural right" (26).
23. See Lincoln (1863).
24. This argument from the equality necessitated by the American democracy is to be found in Grisez and Boyle (1979, 25–34).
25. Giving adequate pain relief even at dosages that decrease life span is permissible under the Principle of Double Effect as well as under the American Medical Association's traditional distinction between ordinary and extra-ordinary means of treatment as long as the death of the patient is not intended.

References

Boyle, J.M. 1979. That the fetus should be considered a legal person. *The American Journal of Jurisprudence* 59:59–71.

Boyle, J.M. 1980. Freedom, Christian values and secular values. In *The battle for the Catholic mind: Catholic faith and Catholic intellect in the work of the fellowship of Catholic scholars 1978–1995*, eds. W.E. May and K.D. Whitehead, 181–205. South Bend: St. Augustine's Press.

Committee of the Harvard Medical School to Examine the Definition of Brain Death. 1968. Report of the Ad Hoc committee of the Harvard Medical School to examine the definition of brain death. *Journal of the American Medical Association* 205:85–88. Reprinted In *Bioethics: An anthology*, eds. P. Singer and H. Kuhse, 287–291. Malden, MA: Blackwell.

Fox News.com. 2008. Transgender woman sues Catholic hospital for refusing breast enlargement surgery. http://www.foxnews.com/story/0,2933,321746,00.html. Accessed 19 Jan 2008.

Frankl, V. 1946. *Ein Psycholog erlebt das Konzentrationslager.* Published in English (1959) as *Man's search for meaning*. New York, NY: Pocket Books.

Fuller, L. 1969. *The morality of law*, Rev. ed. New Haven and London: Yale University Press.

George, R.P. 2002. *The clash of orthodoxies*. Washington, DC: ISI Books.

Grisez, G., and Boyle, J. 1979. *Life and death with liberty and justice: A contribution to the euthanasia debate*. Notre Dame and London: University of Notre Dame Press.

Hayden, R.M. 1989. The paradox of Aquinas' altruism: From self-love to love of others. *Proceedings of the American Catholic Philosophical Quarterly* 63:72–84.

Hayden, R.M. 1991. Aquinas and natural human fulfillment: Inconsistencies? *Proceedings of the American Catholic Philosophical Quarterly* 65:215–233.

Henig, R.M. 2005. At war with their bodies, they seek to sever limbs. The New York Times. http://www.nytimes.com/2005/03/22/health/psychology/22ampu.html?_r=1&adxnnl=1&oref=slogin&adxnnlx=1258560034-cZLs7gFh4f0Tvxky9RCyJA. Accessed 22 Mar 2005.

Kant, I. 1993. *Grounding for the metaphysics of morals with on a supposed right to lie because of philanthropic concerns, 3rd edition.* (trans: Ellington, J.W.). Indianapolis and Cambridge: Hackett Publishing Company.

Lavery, J.V., Boyle, J., Dickens, B.M., Maclean, H., and Singer, P.A. 2001. Origins of the desire for euthanasia and assisted suicide in people with HIV-1 or AIDS: A qualitative study. *The Lancet* 358:362–367.

Lemmons, R.M. 2010. Does suffering defeat eudaimonic practical reasoning? *Proceedings of the American Catholic Philosophical Quarterly* 83:155–172.

Lincoln, A. 1863. Gettysburg address in American historical documents: 1000–1904. In *The Harvard Classics*, Ed. C. W. Eliot, p. 441. New York, NY: P. F. Collier & Son Company.

Overdorf, J. 2008. Stolen kidneys. Newsweek. http://www.newsweek.com/id/107448. Accessed 1 Feb 2008.

Pakaluk, Michael. 1994. Political friendship. In *The changing face of friendship*, ed. Leroy S. Rouner, 197–212. Notre Dame: University of Notre Dame.
Singer, P. 1993. *Practical ethics, 2nd edition*. Cambridge, MA: Cambridge University Press.
St. Thomas Aquinas. 1947. *Summa theologiae*. New York, NY: Benzinger Brothers.
Sudworth, J. 2006. Inquiry into 'Beggar Amputations.' BBC News. http://news.bbc.co.uk/2/hi/south_asia/5231022.stm. Accessed, 31 Jul 2006.
Thomson, J.J. 1977. A defense of abortion. *Philosophy and Public Affairs* 1:47–66. Reprinted in *Bioethics: An anthology*, eds. P. Singer and H. Kuhse, 21–35. Malden, MA: Blackwell.
Tooley, M. 1972. Abortion and infanticide. *Philosophy and Public Affairs* 2:37–65. Reprinted in *Bioethics: An anthology*, eds. P. Singer and H. Kuhse, 21–35. Malden, MA: Blackwell.
Wojtyla, K. 1961. Personalizm tomistyczny. *Znak* 13:664–675. Reprinted as Thomistic personalism. In Wojtyla, K. 1993. *Person and community*. (trans: Sandok, T. OSM), 165–175. New York, NY: Peter Lang.
Wojtyla, K. 1993. *Love and responsibility*. (trans: Willetts, H.T.). San Francisco, CA: Ignatius Press.
Woodward, C. 2007. Pharmacists sue over WA emergency contraception rules. The Seattle Times. http://seattletimes.nwsource.com/html/localnews/2003807209_webpharmacists27m.html. Accessed 26 Jul 2007.
Zwillich, T. 2005. Bill says pharmacies must fill birth control Rx. WebMD. http://www.webmd.com/sex/birth-control/news/20050414/bill-says-pharmacies-must-fill-birth-control-rx. Accessed Apr 14 2005.

Chapter 5
The Significance of the Ultimate End for the Feeding of PVS Patients: A Reply to Kevin O'Rourke

Peter F. Ryan, S.J.

Fr. Kevin O'Rourke argues: "If a person does not have the ability nor the potency to perform human acts now or in the future, then that person can no longer strive for the purpose of human life and it does not benefit the person in this condition to have life prolonged." He concludes that "there is no moral obligation to prolong the life of persons in vegetative state from which they most likely will not recover," and thus holds that caregivers need not arrange for the feeding and hydration of such patients (O'Rourke, 2008a, 174).

In an allocution to participants in the international congress on "Life-Sustaining Treatments and Vegetative State: Scientific Advances and Ethical Dilemmas" on March 20, 2004, Pope John Paul II (John Paul II, 2004, no. 2) defined the so-called vegetative state as one in which the patient "shows no evident sign of self-awareness or of awareness of the environment, and seems unable to interact with others or to react to specific stimuli." To this definition O'Rourke adds that such patients display "sleep-wake cycles; hence, the patients [*sic*] eyes are often open, but unable to track in a meaningful manner" (2008a, 168).

O'Rourke's claim that there is no moral obligation to feed PVS patients contradicts the teaching of the Roman Catholic Church, most recently articulated in John Paul II's allocution and in the Congregation for the Doctrine of the Faith's *Responses to Certain Questions of the United States Conference of Catholic Bishops Concerning Artificial Nutrition and Hydration* (ANH) and the accompanying *Commentary* of September 16, 2007.[1] Moreover, it is questionable whether the vegetative state can, as O'Rourke maintains, be reliably diagnosed as permanent (PVS).[2] But O'Rourke's claim that PVS patients need not be fed would contradict the Church's teaching in those documents even if there were no dispute about diagnosis. So also, my arguments would not be affected even if PVS could be diagnosed with complete certainty. Therefore, I will not consider that controversy here.

I propose to show that O'Rourke unsoundly argues that there is no moral obligation to provide food and water to persons who will never be capable of human acts

P.F. Ryan, S.J. (✉)
Mount St. Mary's Seminary, Emmitsburg, MD, USA
e-mail: pryan@msmary.edu

C. Tollefsen (ed.), *Bioethics with Liberty and Justice*, Philosophy and Medicine 110,
DOI 10.1007/978-90-481-9791-0_5, © Springer Science+Business Media B.V. 2011

because, being unable to pursue the spiritual purpose of life, they cannot be benefitted by having their lives prolonged. I will argue that O'Rourke has an erroneous notion of the ultimate end and how it should shape our action.

This article has four parts. The first sets out the premises—O'Rourke's view of the ultimate end and how it should shape action—from which he draws the conclusion that there is no moral obligation to feed PVS patients. If those premises were true, the conclusion would follow. So, the second part argues that the premises are false. The third part proposes an alternative understanding of the ultimate end and how it should shape action. The fourth part applies that alternative understanding to the question of feeding PVS patients and concludes that a proper consideration of benefits and burdens almost always reveals a moral obligation to feed them.

5.1 O'Rourke's View of the Ultimate End and How It Should Shape Action

O'Rourke clearly identifies the issue that underlies his dissent from the teaching that caregivers are normally obliged to provide ANH for patients in PVS: "The positive reasons for disagreement with the teaching contained in the allocution are founded upon a Thomistic anthropology of the human person" (O'Rourke, 2008a, 174), and more specifically "upon an interpretation of the Thomistic concept of the purpose of human life" (O'Rourke, 2008b, 190). To come to grips with that purpose, we must consider the teaching of St. Thomas Aquinas on the ultimate end and its pursuit.

5.1.1 St. Thomas on Pursuit of the Ultimate End

According to Thomas, human beings naturally desire absolutely perfect fulfillment, and necessarily take as their ultimate end something that they expect to be perfectly fulfilling.[3] He explains: "Since everything desires its own perfection, what one desires as an ultimate end, one desires as a perfect good that completes oneself." And, since a perfect good can leave nothing to desire, one cannot at any one time have more than one ultimate end: "It is therefore necessary that an ultimate end so fill a human being's entire appetite that nothing else is left to desire. That is not possible if something more were required for one's perfection. Therefore, it is not possible for the appetite to tend to two things as though each were its perfect good" (Aquinas, *ST*, I–II, Q. 1, a. 5). Several points in this teaching require clarification.

First, for Thomas, although we can change what we take to be perfectly fulfilling, we cannot at the same time take two different things to be the good that will perfectly fulfill us. So, a person cannot make successive choices for the sake of different ultimate ends unless he repudiates in the subsequent choice the ultimate end for whose sake he made the previous choice. Thus, for example, a man cannot, on Thomas's view, do household chores for the sake of his marriage, and order that activity to God as his ultimate end, and then sleep with his mistress and order that activity to that enjoyable relationship as an end he desires for its own sake, without repudiating God as his ultimate end.

Second, Thomas recognizes that one can take an aggregate of goods as one's ultimate end—not if each is desired for its own sake alone, but only if they are desired as aspects of a complex good regarded as perfectly fulfilling (see Aquinas, *ST*, I–II, Q. 1, a. 5, ad 1).

Third, Thomas is not just saying that one necessarily seeks the *ratio*, or idea of perfect fulfillment—perfect fulfillment in general—but that one necessarily takes something or other as the perfectly fulfilling good. This is clear because he says that one's will can only be directed to one thing and not to "*diverse things* as ultimate ends" (Aquinas, *ST*, I–II, Q. 1, a. 5, my emphasis). That statement makes it clear that he has in mind different things in which one might think the *ratio* of the ultimate end is found (see Grisez, 2008, 41).

Thomas teaches that "a human being necessarily desires everything he desires for the sake of an ultimate end" (Aquinas, *ST*, I–II, Q. 1, a. 6). He adds, however, that one need not be thinking of an ultimate end whenever one desires or acts for something, for "the power of the first intention, which is in respect of an ultimate end, remains in every desire for any object whatever, even if one is not actually thinking about the ultimate end" (Aquinas, *ST*, I–II, Q. 1, a. 6, ad 3). This explanation implies that one's first intention—one's intention of the ultimate end—is conscious, at least initially. By way of analogy, Thomas points out that one need not be thinking of one's destination at every stage of a journey—but, of course, one can only begin the journey by having the destination in mind.

Thomas also holds that at any given time, one necessarily directs all one's choices and, therefore, all one's human acts, to some single ultimate end. This point is entailed by his teaching that one can have only one ultimate end at a time, and that one necessarily wills whatever one wills for the sake of some ultimate end.

Finally, Thomas makes it clear that what one takes to be one's ultimate end is not a given. He explains: not all agree on where perfect fulfillment is to be found; we seek it in different things; and "by sinning, some turn away from the unchangeable good" (Aquinas, *ST*, I–II, Q. 1, a. 7, obj. 1), which is the true ultimate end. Thus he indicates that we are responsible for what we take as our ultimate end. It is worth noting that since a person can, on this view, have only one ultimate end at a time, one will be completely off course unless what one takes as one's ultimate end is the true ultimate end. Conversely, the person who does take as ultimate end the true ultimate end will, by doing so, completely organize his or her life. For, again, on this account, all one's choices and human acts are directed to what one takes, rightly or wrongly, as one's single ultimate end.

5.1.2 St. Thomas on the Beatific Vision as the True Ultimate End

Having argued that we necessarily seek absolutely perfect fulfillment, and that we necessarily do so by pursuing something as our single ultimate end, Thomas claims that attaining God by the beatific vision is the end we should pursue, the true ultimate end. He says that the intellect naturally desires to know the essences of the causes of the effects it knows, and ultimately desires to know the essence of the first cause, which is God himself (see Aquinas, *ST*, I–II, 1, Q. 3, a. 8). Thomas implies that nothing other than the beatific vision can fulfill the natural desire for perfect happiness, since anything less would leave unfulfilled the intellect's natural desire to know the essence of the first cause.

The reason the beatific vision is so completely satisfying, according to Thomas, is that in it one actually attains divine goodness: "Man's mind will be united to God by one, continual, everlasting operation" (Aquinas, 1920, I–II, Q. 3, a. 2, ad 4). Since this operation is of one's "highest power," the intellect, "in respect of its highest object ... the Divine Good" (Aquinas, 1920, I–II, Q. 3, a. 5), Thomas concludes that the beatific vision is one's best possible operation. In it, the divine essence becomes the very form of the intellect's act.

Thomas explains that in this vision, one's "desires are completely brought to rest." He says the vision includes "the full sufficiency of all the goods which, according to Aristotle, are required for happiness" and thus also meets Boethius' definition of happiness as "a state of life made perfect by the accumulation of all goods" (Aquinas, 1955–1957, III, 63, 9; cf. Boethius, *De consolatione philosophiae* III, 2). Thomas affirms that "whatever good there be in these [other] things, we shall possess it all in the Supreme Fount of goodness" (Aquinas, 1920, I–II, Q. 4, a. 7, ad 2).

Thomas also teaches that everyone who has the beatific vision is completely fulfilled, but that people have it to varying degrees, according to their capacity (see Aquinas, *ST*, I–II, Q. 5, a. 2; Aquinas, 1955–1957, III, 59, 4). He explains that one's capacity depends on one's charity. The blessed who have more charity "will see God the more perfectly, and will be the more beatified" (Aquinas, 1920, I, Q. 12, a. 6). Through one's acts here and now, one can increase one's charity and thus one's capacity for that vision: "Each act of charity disposes to an increase of charity," and greater acts of charity actually increase one's charity (1920, II–II, q. 24, a. 6, c).

Thomas considers what goods the blessed have besides the beatific vision, including the resurrection of the body and the community of friends, and discusses how those blessings are related to the vision. Since "perfect Happiness consists in the vision of the Divine Essence" (Aquinas, 1920, I–II, Q. 4, a. 5), those blessings are not essential to beatitude. That the resurrection of the body is not essential is, for Thomas, clear from the fact that many have the beatific vision before they are raised. The divine essence, in any case, "cannot be seen by means of phantasms" (Ibid.). As for the community of friends, Thomas argues that our perfect happiness consists in God alone, and that it is therefore possible in principle for one soul to be perfectly happy without anyone else (see Aquinas, *ST*, I–II, Q. 4, a. 8, ad 3). Nevertheless, Thomas argues that bodily resurrection and the community of friends add to the "well being" of perfect happiness and participate in it by a certain overflow from the beatific vision.

5.1.3 O'Rourke's View of How the Ultimate End Should Shape Action and Its Significance for the Feeding of PVS Patients

Presupposing the Thomistic account, O'Rourke holds that the ultimate end requires actions that are a means to attaining it, and he assumes that no other kinds of actions have any value. His explanation of why ANH need not be provided for patients in PVS makes that clear: Nutrition and hydration may be withdrawn "because such a person can no longer perform acts that will serve the ultimate end of human life, union with God" (Ashley et al., 2006, 194).

The ultimate end, on this view, is a goal extrinsic to the goods realized in or by actions done to attain it. O'Rourke explains: "Although earthly human life is a great good which makes possible our striving for other goods . . . it is only a way to eternal life, from which it takes its full meaning" (Ashley and O'Rourke, 1997, 425). Any increase in our capacity for the beatific vision, brought about by our greater acts of charity, will remain, but the means to attain eternal life, being extrinsic to eternal life itself, will fall away. Of course, the body will be raised up again, and the beatific vision will be enjoyed in the company of friends, but, as we have seen, on this view those blessings are not essential to beatitude.

On this account of the ultimate end and how it should shape action, O'Rourke's conclusion regarding the feeding of PVS patients follows. To support his account and conclusion, however, he quotes a statement by Pope Pius XII to a group of physicians:

> Normally one is held to use only ordinary means [to sustain life]—according to the circumstances of persons, places, times, and cultures—that is to say, means that do not involve any grave burdens for oneself or for another. A more strict obligation would be too burdensome for most people and would render the attainment of higher, more important good too difficult. *Life, health, all temporal activities are in fact subordinated to spiritual ends* (Pius XII, 1958).

O'Rourke interprets that statement as follows: "In other words, the means to prolong life may be withheld or withdrawn if these means do not help a person strive for the spiritual purpose of life, or if they impose a grave burden on the person or the person's caregivers in regard to striving for that purpose" (Ashley and O'Rourke, 1997, 425–426; Pius XII, 1958).

O'Rourke then says that prolonging life "is of benefit only when it gives the person opportunity to continue to strive to achieve the spiritual purpose of life" (Ashley and O'Rourke, 1997, 426). Since such striving requires "some degree of cognitive-affective function," which, O'Rourke assumes, patients diagnosed as being in PVS have permanently lost, they never will be able to act for their ultimate end. They will never be able to grow in charity and so increase their capacity for fulfillment in the beatific vision. Therefore, they cannot benefit from being fed. O'Rourke explains:

> As long as any consciousness is possible to us, even in a state of suffering, we can perform acts of faith, of love, and of hope that will bring us closer to God and our neighbor, but when

> this becomes no longer possible because consciousness and freedom have been irrevocably lost by deterioration of the brain, the artificial prolongation of life by medical techniques used merely to demonstrate human ingenuity ceases to be of any real benefit to the subject (Ashley and O'Rourke, 1997, 426).

If ANH cannot in any way benefit the patient in PVS, then providing it is pointless and thus not morally required.

But O'Rourke does not claim only that ANH cannot benefit patients in PVS; he claims that providing ANH is seriously burdensome because it entails the burden of total patient care. He says: "When one decides to provide life support in such cases, one is also creating a situation in which one is responsible for the total care of the patient. One must, therefore, take into account not only the burdens of such a procedure as starting an IVF but the burdens of total care, which may lead to the neglect of one's other responsibilities" (Ashley and O'Rourke, 1997, 421–422). O'Rourke also mentions other burdens: "Furthermore, one must take account of the fact that the indignity of existing in a state of persistent cognitive-affective deprivation can be counted as a serious burden to the patient, even if the patient is unconscious, along with the burdens the family may endure in caring for a person in such a condition" (Ashley and O'Rourke, 1997, 421–422; also see O'Rourke, 2008b, 190–191).

Since on this account there is not only no benefit but also a great burden, then a fortiori one must conclude that feeding PVS patients is an extraordinary means and not morally required. Indeed, though I have not found O'Rourke making this point, on his account caregivers could not rightly provide ANH, since it is wasteful to do what cannot benefit the patient.

Finally, O'Rourke acknowledges that withholding nutrition and hydration from PVS patients would be euthanasia if one intended to cause death, but he claims that "if the intention of the act (*finis operis*) is simply 'letting die,' it is not euthanasia." (Ashley et al., 2006, 195).

5.2 Critique of O'Rourke's View of the Ultimate End and How It Should Shape Action

As we have seen, O'Rourke presupposes the Thomistic account of the ultimate end. That account holds (1) that we are inherently oriented to absolutely perfect fulfillment; (2) that we necessarily pursue something definite as our single ultimate end, and (3) that attaining God alone by the beatific vision is the end we should pursue, the true ultimate end, because it alone can completely fulfill us. In this section I first respond to each of those claims, and then criticize O'Rourke's understanding of how the ultimate end should shape action and its significance for the feeding of PVS patients.

5.2.1 Do We Necessarily Seek Absolutely Perfect Fulfillment?

Our experience of making choices shows that we do not seek absolutely perfect fulfillment. In deliberating about options for choice, we know that we must give up something that we will never have the chance to do again. For example, a mother chooses to respond to an emergency at work that she knows will prevent her from being present at her son's high school graduation. Though the son may understand, and the mother may mitigate the deprivation by showing her love in some other way, she never will be able to go to that graduation.

People give up part of what they regard as fulfilling even in the choices by which they organize their lives. For example, a mafia boss may have wanted the good opinion and friendship of decent people he admired, but he gave up those things in order to get other things he wanted. Another man wants the riches that would be available to him though a life of crime, but declines that life and its benefits in order to pursue the goods of an upright life. Each knew he could not have everything he wanted, and made choices that he knew were limiting in order to pursue the fulfillment he preferred.

As young people are growing up, they have to make career choices. This usually means giving up, with a sense of regret, some good things in order to settle on other things. Right-minded people make commitments: They deliberate about morally acceptable options, choose one, and remain true to their choice. In our cultural context, however, people often do not want to make commitments. They want to "have it all," which requires avoiding commitments so they can get as much as possible. From their experience of making choices they know that no definite set of things they might live for will completely satisfy them, but they want to be as fulfilled as they can be. So, to ensure that they will always be free to try something else later, they avoid commitments. To mitigate the limitations of the current bundle of things they seek, they keep their options open and avoid getting locked in, knowing that there are likely to be other things they will want to seek.

Insofar as those who adopt this approach avoid commitments and fail to value the fulfillment of people other than themselves and those near and dear, their approach is deeply flawed. Nevertheless, the approach makes sense insofar as those who adopt it recognize that many things are worthwhile, see the point in being all they can be, and do not seek fulfillment haphazardly or pursue the illusion of absolutely perfect fulfillment.

Thus, it is clear that people do not seek absolutely perfect fulfillment—not just because in choosing certain things, everyone definitively excludes some other things, but also because many people include within the bundle of the things they value the freedom to seek still other things. That they value not getting locked in shows that they do not regard the set of things they currently live for as perfectly satisfying. In order to get as much fulfillment as they can, they have to exclude viewing their ultimate end as perfectly fulfilling.

St. Thomas realized that people can pursue an aggregate of things as ultimate end, but he thought of it as a group whose elements are an intelligible whole regarded as perfectly fulfilling. Today people think of the aggregate as the bundle of things they happen to want, and they do not think of that whole bundle of things as perfectly fulfilling. The evidence is that, being unwilling to commit themselves to only these things, they keep their options open—not because they think they might eventually happen upon a bundle that they think will be perfectly fulfilling, but because they expect they will eventually want things other than at least some of those they now want.

5.2.2 *Must the Acting Person Have a Single Ultimate End?*

If, as Thomas teaches, we necessarily seek absolutely perfect fulfillment in something definite that we regard as leaving nothing to desire, then his conclusion that at any one time a person can have but a single ultimate end inevitably follows.[4] However, if that conclusion is false (if a person can have more than one ultimate end at a time) then the premise also is false (one need not seek perfect fulfillment that leaves nothing to desire). I will show that we can have more than one ultimate end at a time, and thus provide another reason for denying that we are oriented to absolutely perfect fulfillment.

If the acting person can have only one ultimate end at a time, two highly implausible conclusions follow. First, all those who are in the state of grace—who, according to Thomas, have God as their single ultimate end—order all their diverse good acts to God. Second, mortal sinners do everything, including disparate bad deeds and apparently good deeds, for the same ultimate end.

The former conclusion is implausible because some people in grace have never even thought about having a unified life plan and doing all the good things they do for God as their single ultimate end. Yet, as we have seen, one cannot adopt an ultimate end without being conscious of doing so. Moreover, if everyone in the state of grace had God as their single ultimate end, their lives would be completely organized. But then there would be no need for the Church to call Christ's faithful people to holiness by inviting them to order their whole lives to God. Yet such exhortation, which comprises part of the program of the Catholic Church according to Vatican II, plainly is very much needed.

The conclusion in the case of mortal sinners has two levels of implausibility. First, in seeking what they want, mortal sinners not infrequently act at cross purposes. For example, a corrupt politician may curry favor with supporters by cracking down on prostitution and at the same time jeopardize his political position by visiting a prostitute. It is hard to imagine how such acts could be done for the same ultimate end. Second, sometimes mortal sinners do good acts for good reasons. For example, a divorced Catholic man who marries outside the Church, realizing that doing so is gravely wrong, may continue to go to Mass and pray, hoping for salvation. It is especially hard to imagine how those acts could be done for the same ultimate end.

Someone might respond by saying that the single ultimate end of the people in all of the above cases is the aggregate of the things they want. However, as noted above, Thomas understood an aggregate as a group whose elements are an intelligible whole regarded as perfectly fulfilling—and it is clear that people need not think of the various things they seek in that way. For example, many people who live in sin do not expect perfect fulfillment and rationalize their sinful lifestyles by arguing that their best option is to seek whatever fulfillment is available here and now in the diverse things they happen to want.

A final argument against Thomas's view is that the deliberate venial sins of those in the state of grace must be done for some ultimate end other than God. This argument shows that the claim that the acting person must have a single ultimate end is not just implausible, but utterly untenable. According to Thomas, the single ultimate end of a person in the state of grace is God, meaning that the person's every act is done for the sake of union with God in the beatific vision. If this were true, then even the deliberate venial sins of a person in the state of grace would be done for the sake of union with God, since, as Thomas rightly notes, every human act is done for the sake of some ultimate end. However, it is impossible for a person to do something that he or she realizes is offensive to God—even if it does not deprive the person of grace—for the sake of being united with God.

Thomas tries to address the problem by teaching that the order of the venial sin to God is not actual but only habitual (see Aquinas, *ST*, I–II, Q. 88, a. 1, ad 3). But even if an act is only habitually ordered to an ultimate end, it must be ordered to it per se. The act must be done for the sake of the ultimate end even if one does not consciously advert to that end. However, nothing can be done for the sake of an end unless it is thought to be at least possibly conducive to that end—and deliberate venial sinners know that their act cannot be conducive to union with God. It must, therefore, be done for some other ultimate end.

Thus, the person in the state of grace who commits a deliberate venial sin has at least two ultimate ends: the end for the sake of which he deliberately sins venially, which cannot be God, and God, for whose sake he does some and perhaps all of the other things he does.

Since the acting person can have more than one ultimate end at a time, it follows that human beings are not oriented to absolutely perfect fulfillment.

5.2.3 Is the Beatific Vision Alone the True Ultimate End?

Thomas identifies the beatific vision as the true ultimate end, because it alone can satisfy our natural desire for perfect fulfillment.[5] However, I have shown that we have no such desire. But if we have no such desire, it is not reasonable to argue that the beatific vision is the true ultimate end because it alone can satisfy it.

But now a further question arises: Does the beatific vision really satisfy all desire, such that it is impossible for those who enjoy it to desire anything more? That question must be answered in the negative, as the following considerations show.

Catholic theologians, including St. Thomas, have traditionally agreed that Jesus had the beatific vision during his entire earthly life (see Aquinas, *ST*, III, Q. 9, a. 2). Yet he clearly had desires, as his wholehearted commitment to preaching the gospel shows. He desired to do the Father's will, no matter what the cost to himself, because he wanted his hearers to repent and believe the Good News: "I came to cast fire upon the earth; and would that it were already kindled!" (Lk 12:49).

Jesus not only had desires during his earthly life; he continues to have desires even now. The Letter to the Hebrews (7:25) tells us that "he is able for all time to save those who draw near to God through him, since he always lives to make intercession for them," and Paul confirms that Jesus "intercedes for us" (Rom 8:34). Jesus could not intercede without desiring the salvation of those for whom he intercedes. This is also true of Mary: She, too, intercedes for us, and could not do so without desiring that we receive the benefits she seeks on our behalf. Yet both Jesus and Mary have the beatific vision. Therefore, the beatific vision does not satisfy all desire.

Again, as we have seen, Thomas holds that "perfect happiness consists in the vision of the divine essence," which some enjoy even before rising from the dead. But if all their desires are satisfied without resurrection, they cannot desire it, and then the question arises: What difference can it make to their happiness? Realizing that bodily resurrection is a central truth of faith, Thomas tries to resolve this problem by claiming that the body, though not "necessary for man's perfection," is "necessary for its well-being" (Aquinas, 1920, I–II, Q. 4, a. 5). However, this claim leaves one wondering how, if the beatific vision perfectly fulfills the human person, resurrection can contribute to his or her well being.

Thomas's conviction that the beatific vision satisfies all desire also leads him to claim that it is possible in principle for one soul to be perfectly happy without any friends (see Aquinas, *ST*, I–II, Q. 4, a. 8, ad 3). This claim, too, raises a question: What difference can that community possibly make to heavenly happiness? Thomas again argues that it adds "well-being" to perfect happiness (see Aquinas, *ST*, I–II, Q. 4, a. 8), but one is again left wondering how, if beatitude really is perfect, anything could make life better for those who enjoy it.

In sum, one cannot explain how bodily resurrection and the community of friends contribute anything to our heavenly happiness, as Thomas plainly is convinced they do, without denying that the beatific vision satisfies all desire. But if those who have the beatific vision can have other desires, then the true ultimate end cannot be the beatific vision alone, and must include the objects of those other desires.

5.2.4 Critique of O'Rourke's View of How the Ultimate End Should Shape Action and Its Significance for the Feeding of PVS Patients

As noted above, O'Rourke holds that the ultimate end calls for actions that are a means to it, and he assumes that no other actions have any value. However, if the ultimate end is God attained by the beatific vision, it is not a goal or state of affairs toward which one can work, for we can do nothing to bring it about. No doubt, O'Rourke would readily affirm that the beatific vision is a gift of grace. Still, his understanding of how the ultimate end should shape action does not adequately take that truth into account.

As for the resurrection of the body and the community of friends, even if O'Rourke agreed that they are necessary for beatitude, he would not conclude that the value of PVS patients' bodily life and of communal solidarity with them in this world warrants feeding them, for his view of how the ultimate end should shape action would preclude such a conclusion. He assumes that actions have value only insofar as they help one attain the ultimate end, and therefore holds that because PVS patients cannot pursue that end, there is no point in feeding them. Moreover, he thinks that people's actions now are at best only means to attain the ultimate end—means that will fall away—and that the ultimate end is extrinsic to the goods realized in or by those actions. He therefore concludes that actions to promote the bodily life of PVS patients and communal solidarity with them do not contribute to the ultimate end and so cannot justify feeding them.

This extrinsicist view is evident in Thomas's claim that bodily fulfillment and fulfillment in the community of friends follow upon the beatific vision. As noted above, he provides for continuity between what we do now to increase charity and our capacity hereafter for fulfillment in the beatific vision. But Thomas does not explain how promoting other goods here and now is related to our fulfillment in those goods hereafter. He does not account for continuity with respect to those goods, for he treats fulfillment in them hereafter—which ultimately is fulfillment in resurrection life and the human goods it includes—as a fresh reality deriving entirely from the beatific vision. O'Rourke's comment that "earthly human life . . . is only a way to eternal life, from which it takes its full meaning" (Ashley and O'Rourke, 1997, 425) makes it clear that he agrees. On this view, our fulfillment in bodily life and human community in heaven is brought about by an overflow from the beatific vision, and thus is different in kind from the fulfillment we have by being alive, healthy, and having friends in our present life.

O'Rourke's interpretation of Pius XII's statement that *"Life, health, all temporal activities are in fact subordinated to spiritual ends*" also merits criticism. The Pope neither said nor implied that "the means to prolong life may be withheld or withdrawn if these means do not help a person strive for the spiritual purpose of life" (Ashley and O'Rourke, 1997, 425–426; Pius XII, 1958). He was teaching that, although caregivers have an obligation to preserve life and health, they must not ignore more important spiritual values, and therefore the obligation does not extend to the provision of excessively burdensome care. But Pius neither spelled out what it means to attend to spiritual values nor articulated a principle for determining when care is excessively burdensome (see Congregation for the Doctrine of the Faith, 2007b).

As we have seen, O'Rourke's conclusion about not feeding PVS patients follows from his premises about the ultimate end and how it shapes action. His principle for doing or omitting something is whether it will help a person reach the ultimate end of the beatific vision. Another reason for criticizing that principle is that an obviously absurd conclusion follows from it. Since persons who will never be capable of a human act after they are baptized are assured of the beatific vision, it follows that killing such persons right after they are baptized would be an act of charity since it would mean helping them reach that vision right away.

Of course, O'Rourke would firmly reject such a conclusion and the proportionalist reasoning it presupposes (see Ashley and O'Rourke, 1997, 411–421). He says:

"Killing an innocent person is never an ethically acceptable remote or proximate intention. Human life is a basic good, and we should never directly act in opposition to basic goods" (O'Rourke, 1996, 2). However, human life cannot be a basic good if people are no better off alive than dead—and that would be true of PVS patients if they cannot be benefitted by being fed. And if human life is not a basic good, there is no reason to think that intentionally killing innocent persons—including euthanizing PVS patients—is never ethically acceptable.

Still, O'Rourke staunchly denies that he is advocating euthanasia. He says that it would not be euthanasia to withhold nutrition and hydration from PVS patients in ordinary circumstances if one's proximate intention is not to cause death but only "letting die"[6] (Ashley et al., 2006, 195). However, for two reasons that distinction cannot be sustained. First, as Chris Tollefsen (2008, 214–215) explains, one *may* accept a foreseen death as a side effect only if one chooses to do or omit something else "so as to obtain some benefit." Letting someone die is not itself a benefit. Second, one *can* accept a foreseen death as a side effect only if it truly is a side effect, but O'Rourke treats "letting die" not as a side effect but as the object chosen. As Tollefsen points out, "the whole purpose of the intend-foresee distinction was to make the point that the permitted effects were not part of the agent's intention at all" (2008, 215). On O'Rourke's account, however, the effect of death clearly would be intended.

O'Rourke might reply that he meant to say that the benefit intended could be that of avoiding excessive burdens. If burdens really are excessive, then it is indeed morally legitimate to withdraw ANH in order to avoid them, and to accept the side effect of death. But O'Rourke has not shown that, in ordinary circumstances, it is excessively burdensome to provide ANH to PVS patients. Of course, if, as he claims, feeding PVS patients does not benefit them, then it is by definition excessively burdensome to provide ANH. However, as we have seen, O'Rourke does not vindicate this claim.

O'Rourke attempts to strengthen his case by arguing that providing ANH is a great burden because "one is also creating a situation in which one is responsible for the total care of the patient" (Ashley and O'Rourke, 1997, 421). But if providing a room, a bed, and general nursing care for the patient is the burdensome thing that justifies withdrawing ANH and letting people die, it follows that many people in nursing homes—for example, people suffering from severe dementia whose care is being paid for by Medicaid—who do not need ANH could be rightly denied care and left to die.

The burdens of ANH are one thing and the burdens of the other elements of the patient's care are another, and the two should not be confused. Caregivers may forgo only what is excessively burdensome. ANH may be withheld from the PVS patient only if providing it is in and of itself excessively burdensome—and O'Rourke has not shown that it ordinarily is. The same holds for every other element of patient care: It may be withheld only if providing it is excessively burdensome.

Thus, if a family cannot reasonably afford round-the-clock nursing care in a professional facility, they may avoid that excessive burden by making arrangements for their loved one to stay at home—and if the patient is in PVS, they must leave the feeding tube intact. At home they may avoid excessive burdens by doing only what they reasonably can, given their limited resources and other responsibilities, to provide for the patient's basic needs. But they may not withhold ANH in order to avoid other burdens, for in doing so they would not be accepting the patient's death as a side effect, but choosing the patient's death as a means of avoiding those other burdens.[7]

5.3 An Alternative Understanding of the Ultimate End and How It Should Shape Action

If, as I have argued, the beatific vision alone is not the true ultimate end, the question arises: What is? I respond: The ultimate end is the divine-human community—the kingdom of God—which embraces all persons who live and die in God's friendship, and all goods. The kingdom includes the beatific vision, which is a participation in uncreated goodness—our intimate sharing in divine life, which we are able to share in because at baptism we were made children of God. But the kingdom also is this world transformed into a new creation—the new heavens and the new earth. It includes created, human goods that cannot be reduced to the beatific vision: resurrection life, the communion of friends, and the whole range of humanly fulfilling goods.

How should this ultimate end shape action? Although we cannot bring about the kingdom, we can prepare material for it, confident that the Lord will transform that material and incorporate it into the kingdom, which he will bring about and give to those who enter it as his gift. We should, then, shape our lives not with a view to attaining the kingdom by our actions but with a view to contributing to it by preparing material for it.

Vatican II (1965) supports this understanding of the kingdom and how it should shape action by generalizing the principle of resurrection. After speaking of the call "to make ready the material of the celestial realm" (no. 38), *Gaudium et spes* proclaims that "God is preparing a new dwelling place and a new earth where justice will abide" (no. 39). The pastoral constitution goes on to say: "after we have obeyed the Lord, and in His Spirit nurtured on earth the values of human dignity, brotherhood and freedom, and indeed all the good fruits of our nature and enterprise, we will find them again, but freed of stain, burnished and transfigured" (no. 39). Thus, whereas St. Thomas accounted for continuity only with respect to divine life, *Gaudium et spes* goes further by accounting for continuity with respect to human goods. It teaches that everything—our human acts, our bodily life, and creation itself—is going to be reconstituted and perfected. Thus, earthly human life is not, as O'Rourke claims, "only a way to eternal life" (Ashley and O'Rourke, 1997, 425).

Although we do not inevitably take anything as our single ultimate end, we should take the kingdom, with all of its diverse and manifold goods, as our single ultimate end. Those who do so cannot expect absolutely perfect fulfillment, since nothing can fulfill desire in a way that makes further desire and further fulfillment impossible. But if such persons remain faithful they can expect to be all they can be in the kingdom, not only by sharing in Trinitarian intimacy but also by finding ever-greater happiness in the various goods that are humanly fulfilling.

Unlike modern-day pagans, people who take the kingdom as their single ultimate end see that the path to true fulfillment lies not in pursuing it by avoiding commitments, but in being fully committed to God's plan for their lives, which almost always entails subordinate commitments. They are willing to accept the limitations these commitments involve here and now because they know that only in the kingdom can created persons be all they can be forever. Their fulfillment there will be immeasurably greater than the fulfillment of those who seek it only in this world, because fulfillment in the kingdom is not subject to the limitations of sin and death, and is not individualistic. Fulfillment in the kingdom lasts forever, and it is ever intensified and expanded because members of the heavenly communion rejoice and find fulfillment in the fulfillment of others.

However, those who fail to take the kingdom with all it includes as their single ultimate end are not necessarily utterly off track. They act reasonably to the extent that they promote human goods, and they can be saved as long as they do not consent to anything incompatible with the kingdom. Even if they do not have explicit faith, they can be saved if they accept the grace to live according to what they think is the truth. And, since God incorporates into the kingdom all of the good done by those who enter it, such people prepare material for the kingdom even if they are not aware of doing so.

Consider the example of cultivating solidarity here. Such activity is not just a means to an extrinsic end but is built into that end, for relationships cultivated here continue in heaven.

The traditional assumption, which O'Rourke apparently shares, is that it is because a person has the beatific vision that he or she enjoys other benefits, including fulfillment in the community of friends. On that conception, the communal life of the kingdom is not something that is built up here and now, and continues and deepens in the kingdom. Rather, it is entirely new, for it flows from the beatific vision.

I am proposing, by contrast, that it is because a person enters the kingdom—the everlasting community of divine and human persons—that he or she has the beatific vision. On this conception, the community of friends is being built up even now, and those who enter into the kingdom share intimately in each others' lives in ways that build upon everything humanly good that went before. This communal intimacy also includes the beatific vision, which is the intimate sharing in the divine goodness proper to the divine persons and the maturation of the sharing in divine life that believers received at baptism (1 John 3:2). Sharing in the beatific vision is a consequence of sharing in the good of human friendship with Jesus and bodily incorporation into him.

5.4 Application of the Alternative View to the Problem of Feeding PVS Patients

To understand how all of this applies to the question of whether there is a moral obligation to feed patients in PVS, we must first recognize that one can both harm and benefit people without their being aware of it. For example, ruining someone's reputation through slander harms the person even if that person dies without learning of the harm. So also, a husband who commits adultery does an injustice to his wife—he harms her—even if she never discovers his infidelity.

Conversely, a person is benefitted when truthfully spoken of well by others, even if the person never learns of the benefit. So also, a husband's faithfulness to his wife by resisting the temptation to commit adultery benefits her even if she is unaware that he experienced and resisted the temptation. The benefit—which also is a benefit for him—is in maintaining and strengthening the relationship. People also can be benefitted spiritually without being aware of the benefit—for example, prayers can be offered on their behalf, infants can be baptized, and so forth.

What is true in general is also true of PVS patients: Despite their lack of consciousness, they, too, can be the subjects of harm and benefit. O'Rourke himself says that PVS patients "may be said to have the burden of indignity" (Ashley and O'Rourke, 1997, 424), and thus inadvertently admits that they can be harmed. But these patients can have burdens beyond that of PVS itself. For example, a woman in PVS who is raped is harmed even if she is never aware of the harm. She also can be harmed by slander, her husband's infidelity, and so forth.

So also, those in PVS can be benefitted. A woman in PVS is benefitted when her husband remains faithful to her. The benefit—which also is a benefit for her husband—is in maintaining and strengthening their marital relationship. The provision of food and water also benefits the PVS patient and strengthens the relationship between patient and caregiver. This is so because ANH sustains the patient's bodily life, which is a good of the person, and providing it is faithfully caring for the person, which is good for both parties

O'Rourke agrees that bodily life is a good of the person, but denies that sustaining the bodily life of a PVS patient benefits him or her. If that were true, then killing a PVS patient would not harm him or her, and, pace O'Rourke, would not be wrong and might well be obligatory. Perhaps O'Rourke would respond that although the killing cannot harm the PVS patient, it harms the one who does the killing. But how can killing harm the one who kills if it does not harm the person killed?

In short, the caregiver who does something to improve a PVS patient's bodily well being benefits him or her and treats the person with respect. Providing food and water contributes to the bodily well being of PVS patients and thus benefits them, even if they are never aware of the benefit. And, because providing food and water and other aspects of care also maintains and builds up community, those actions benefit not only PVS patients but also their caregivers.

PVS patients also can be benefitted spiritually, for example, by being given the sacrament of the sick, which customarily is given to any believer, conscious or unconscious, who is even remotely in danger of death. Fr. O'Rourke surely would not say that these patients should not be anointed. If, however, they cannot be benefitted, there is no point in anointing them.

Those who provide food and water for PVS patients prepare material for the kingdom, as one can see by reflection on fulfillment hereafter in the community of friends and bodily life.

Fulfillment in the community of friends will not be a completely fresh reality, as it would be if it were merely an overflow from the beatific vision. Rather it will be in continuity with our relationships here and now. There is no reason to think that we will not recognize each other in the kingdom and build on what went before. What we do for each other here and now is the foundation on which we will be able to carry on ever-deepening relationships hereafter. Those who care properly for PVS patients prepare a radically different foundation from those who do not, and their relationships with those they helped will be forever different than they would be if they did not help them. Such caregivers thus prepare material for the kingdom, and do so even if they never reflect on this truth.

So also, our fulfillment hereafter in bodily life will not be utterly new, a mere overflow from the beatific vision, but rather will be intrinsically connected to our bodily fulfillment here and now, despite the fact that our bodily life in the kingdom will be transformed and all impediments to its ever-deeper fulfillment removed. This point is confirmed by the doctrines of Eucharist (see Lee, 2008, 181) and resurrection. We experience bodily union with Jesus Christ even now in the Eucharist, and that Eucharistic union is a foretaste of bodily union in the kingdom. So also, we are to be raised up in our own body. Since this body will be in the kingdom, what we do for this body now also will be in the kingdom. Whatever good has been done to promote bodily well being here and now will somehow be incorporated into the kingdom with lasting value.

How do these considerations bear on the question of whether there is a moral obligation to feed PVS patients? O'Rourke rightly notes that to judge this matter properly, one must determine whether the benefits of feeding the patient are greater than the burdens. But this obviously requires properly grasping the benefits and burdens.

One can grasp the benefits properly only if one appreciates that the goods at stake, notably the goods of human life and solidarity, are not mere means to the ultimate end but elements of it. Recognizing that truth enables one to see that doing what promotes those goods—in this case, feeding PVS patients—is a significant benefit because it contributes to the ultimate end of both patient and caregiver.

One can grasp the burdens properly only if one distinguishes the burden of feeding PVS patients from the burdens of total patient care. Making that distinction enables one to see that providing ANH is rarely very burdensome in itself. Neither the nasogastric tube nor the gastronomy tube are very costly. Moreover, PVS patients, assuming they are really unconscious, do not experience pain.

A proper analysis of the benefits and burdens, then, makes it clear that normally there is a moral obligation to feed PVS patients. Of course, as O'Rourke emphasizes (Hardt and O'Rourke, 2007) and the CDF statement (2007b) confirms, the teaching that it is morally obligatory "in principle" to administer food and water to PVS patients means that there can be exceptions. Under certain circumstances, ANH can be excessively burdensome, as the statement explains. For example, if the patient cannot assimilate the nutrients and liquids, then they do not sustain the patient's life, and administering them is obviously pointless and should be discontinued (though caregivers are still obliged to do whatever really does help and is not excessively burdensome). So also, the statement notes that "in very remote places or in situations of extreme poverty, the artificial provision of food and water may be physically impossible, and then *ad impossibilia nemo tenetur.*"

However, even if one were to extend "impossible" to cover situations in which ANH cannot be provided without using resources needed to fulfill other exigent responsibilities, it is hard to imagine that there could be many exceptions. As noted above, the burden of ANH must be distinguished from the burdens of total patient care, and in itself ANH is not very costly or labor intensive. And, again, providing nutrition and hydration offers the significant benefits of sustaining bodily life and promoting solidarity, and thus contributes to fulfillment not just here but also hereafter.

In saying that caregivers normally are obliged to administer food and water to PVS patients, I do not mean to minimize the suffering endured by those who faithfully fulfill this obligation. Rather, I wish to call attention to the meaning of that suffering. As Joseph Boyle points out, "If we make good choices we have done our part, and the suffering that ensues because those choices are sometimes of actions that are deeply unsatisfying and troubling is not absurd but justified, though we can only trust that the justification is there" (Boyle, 2008, 117).

Boyle reflects further on this suffering:

> Sometimes we get a hunch or are given a glimpse of that justification, but for many of us, much of the time, the suffering that trying to do what is right leads to is a terrible slog that faith alone can sustain. Keeping people alive who likely will never again experience, think or interact with others seems absurd to many people, but it is not absurd if we think we cannot stop this effort without treating them unfairly, or without breaking faith with them, or even without intentionally ending their lives. The goods realized by refusing these choices are real, and even in our suffering we can experience them (Boyle, 2008, 117).

I offer my reflections in the hope that they will help those who faithfully care for PVS patients appreciate the enduring reality of those goods.

Notes

1. O'Rourke (2008a) responds to John Paul II's allocution; he treats the CDF response in Hardt and O'Rourke (2007).
2. John Paul II (2004, no. 2) points out that "there are well-documented cases of at least partial recovery even after many years." In a statement quoted by the Congregation for the Doctrine of the Faith's *Commentary* (2007b), he says that while the vegetative state itself can be diagnosed, the conclusion that the condition is permanent is not a distinct diagnosis "but only a conventional prognostic judgment, relative to the fact that the recovery of patients, statistically speaking, is ever more difficult as the condition of vegetative state is prolonged in time." O'Rourke argues to the contrary that "the condition of permanent vegetative state can be diagnosed with moral certitude" (2008b, 190; also see O'Rourke 2008a, 168–169). I have argued that the difference of opinion among experts on this matter grounds a reasonable doubt about the actual condition of so-called PVS patients, and concluded that even those who think that only conscious

life is worth living cannot reasonably maintain that ANH is useless for such patients (see Ryan, 2000, p. 48 and n. 16, p. 56; online [1999 contents], p. 8 and n. 16, p. 16).

3. For a fuller treatment of the material treated in this section and the next, see Ryan (2001, part 1).
4. For a fuller treatment of this question, see Ryan (2001, part 2 and following).
5. I am deeply indebted to Germain Grisez for his groundbreaking work on this question. For a much more in-depth treatment, see Grisez (2008).
6. O'Rourke says that Pope John Paul II stated this claim in his "Address to the Participants in the 19th International Conference of the Pontifical Council for Health Pastoral Care," but the Pope neither states nor suggests any such thing.
7. Pat Lee (2008, 184–185) makes this point in his reply to O'Rourke, but O'Rourke (2008b, 190–191) ignores it in his response to Lee. Instead, he answers Lee's observation that "the administration of nutrition and hydration is not expensive or terribly burdensome" by insisting that care for PVS patients is quite burdensome. However, Lee did not claim that other elements of care are not often burdensome, but only that providing ANH itself is not. See also Joseph Boyle (2008, 115–116).

References

Aquinas, T. 1920. *The summa theologica of St. Thomas Aquinas, 2nd and revised edition*. Literally translated by Fathers of the English Dominican Province [On-line]. http://www.newadvent.org/summa/2.htm. Accessed 28 Aug 2010.

Aquinas, T. 1955–1957. *Summa contra gentiles*. (trans: Bourke, Vernon J). Edited, with English updated by Joseph Kenny, O.P. New York, NY: Hanover House [On-line]. www.josephkenny.joyeurs.com/CDtexts/ContraGentiles.htm. Accessed 02 Sep 2010.

Aquinas, T. *Summa theologiae (ST)*. Translations are mine.

Ashley, B., O.P., and K. O'Rourke, O.P. 1997. *Health care ethics: A theological analysis, 4th edition*. Washington, DC: Georgetown University Press.

Ashley, B., O.P., deBlois, J., C.S.J., and K. O'Rourke, O.P. 2006. *Health care ethics: A Catholic theological analysis, 5th edition*. Washington, DC: Georgetown University Press.

Boyle, J. 2008. Towards ethical guidelines for the use of artificial nutrition and hydration. In *Artificial nutrition and hydration: The new Catholic debate*, ed. C. Tollefsen, 111–121. Dordrecht, The Netherlands: Springer.

Congregation for the Doctrine of the Faith. 2007a. *Responses to certain questions of the United States Conference of Catholic bishops concerning artificial nutrition and hydration* [On-line]. www.vatican.va/roman_curia/congregations/cfaith/documents/rc_con_cfaith_doc_20070801_risposte-usa_en.html. Accessed 02 Sep 2010.

Congregation for the Doctrine of the Faith. 2007b. *Commentary* [On-line]. http://www.vatican.va/roman_curia/congregations/cfaith/documents/rc_con_cfaith_doc_20070801_nota-commento_en.html.

Grisez, G. 2008. The true ultimate end of human beings: The kingdom, not god alone. *Theological Studies* 69:38–61.

Hardt, J., and K. O'Rourke, O.P. 2007. Nutrition and hydration: The CDF response, in perspective. *Health Progress* 88:44–47.

John Paul II. 2004. Care for patients in a "Permanent" vegetative state. *Origins 33,* 737: 739–740.

Lee, P. 2008. The papal allocution concerning care for PVS patients: A reply to Fr. O'Rourke. In *Artificial nutrition and hydration: The new Catholic debate*, ed. C. Tollefsen, 179–187. Dordrecht, The Netherlands: Springer.
O'Rourke, K., O.P. 1996. Ethical issues in health care. *Saint Louis University Health Sciences Center, Center for Health Care Ethics* 18(8), 2.
O'Rourke, K., O.P. 2008a. Reflections on the papal allocution concerning care for PVS patients. In *Artificial nutrition and hydration: The new Catholic debate*, ed. C. Tollefsen, 165–177. Dordrecht, The Netherlands: Springer.
O'Rourke, K., O.P. 2008b. Response to Patrick Lee. In *Artificial nutrition and hydration: The new Catholic debate*, ed. C. Tollefsen, 189–191. Dordrecht, The Netherlands: Springer.
Pius XII. 1958. The prolongation of life. *The Pope Speaks* 4:395–398.
Ryan, P., S.J. 2000. The value of life and its bearing on three issues of medical ethics. In *Life and learning IX*, ed. J. Koterski, S.J., 41–57. Washington, DC: Georgetown University. http://www.uffl.org/vol%209/ryan9.pdf. Accessed 02 Sep 2010.
Ryan, P., S.J. 2001. Must the acting person have a single ultimate end? *Gregorianum* 82:325–356.
Tollefsen, C. 2008. Ten errors regarding end of life issues, and especially artificial nutrition and hydration. In *Artificial nutrition and hydration: The new Catholic debate*, ed. C. Tollefsen, 213–226. Dordrecht, The Netherlands: Springer.
Vatican Council II. 1965. *Gaudium et spes*. Vatican City: Libreria Editrice Vaticana.

Part III
Double Effect and Bioethics

Chapter 6
Praeter Intentionem in Aquinas and Issues in Bioethics

E. Christian Brugger

The two propositions in Aquinas from which the so-called doctrine of double-effect arises are (1) actions are morally specified ("receive their species") according to what is intended ("*id quod intenditur*") and not according to what falls outside the intention ("*praeter intentionem*"), and (2) a single act can have effects that fall within the agent's intention and others that fall outside the intention (Aquinas, *ST*, II–II, Q. 64, a. 7). Aquinas famously applies this reasoning to the problem of killing in self-defense and concludes that it is morally licit to kill an aggressor in self-defense provided that one's force is proportioned to the end of self-defense and that the death of the aggressor does not fall within the intention of the agent (i.e., is *praeter intentionem*).

There is considerable disagreement among philosophers as to how the concept of *praeter intentionem* here should be interpreted. Some think it excludes killing an aggressor willed as an end, but permits choosing to kill one as a means to the end of self-preservation (Long, 2007, 39–63). Others think it excludes killing willed both as an end and a means and only permits killing as a side-effect of an act that one foresees *might* happen but must not be foreseen as inevitable (Cavanaugh, 1997). And others think *praeter intentionem* includes unintentional killing not only risked but also foreseen as inevitable (Boyle, 1977, 1978, 1991). Resolving the disagreement is not a mere exercise in scholarly pedantry. A sound interpretation of *praeter intentionem* in Aquinas is necessary for understanding his concept of intention, his wider account of moral responsibility and his general theory of action.

This essay does not address Aquinas's wider theory of action.[1] It only considers the meaning of *praeter intentionem* in Aquinas's treatise on self-defense. It defends the third position above, an interpretation advanced by Joseph Boyle, arguing the following two propositions: (1) effects that fall *praeter intentionem* exclude all effects willed as ends, whether conceived as final purposes (one's reasons for adopting means) or as proximate purposes (the means themselves, conceived as close-in ends beginning with the most proximate); and (2) effects that fall *praeter intentionem* include all effects of an action foreseen or unforeseen that stand outside

E.C. Brugger (✉)
Saint John Theological Seminary, Denver, CO, USA
e-mail: Christian.Brugger@archden.org

C. Tollefsen (ed.), *Bioethics with Liberty and Justice*, Philosophy and Medicine 110,
DOI 10.1007/978-90-481-9791-0_6, © Springer Science+Business Media B.V. 2011

the end-means nexus that constitutes the intelligible plan formulated and acted upon by a moral agent.[2] By way of defending these propositions, the essay sets forth and criticizes two competing accounts each of which respectively denies at least one of these two propositions. The essay ends by applying the concept of *praeter intentionem* to two controversial issues in bioethics.

6.1 Both Ends and Means Intended?

In his text, *The Teleological Grammar of the Moral Act*, Steven A. Long argues that interpretations of Aquinas's argument for licit killing in self-defense have been departing from the real teaching of St. Thomas since the time of Cajetan (d. 1534) (Long, 2007, 29).[3] Long's interpretation of Aquinas's argument can be set forth as follows:

> *Major premise*: the concept of "intention" in Aquinas, when construed according to the "*per se* instance of human action" (a term used by Long to refer to a paradigmatic instance of human action), pertains only to the willing of last ends and not to the willing of means.
>
> *Middle term*: the concept of intention as used in Aquinas's discussion of killing in self-defense (Aquinas, *ST*, II–II, Q. 64, a. 7) should be construed as referring to the "*per se* instance of human action."
>
> *Conclusion*: Aquinas's use of intention in his discussion of killing in self-defense should be construed as excluding only the killing of an aggressor in self-defense adopted as the last end of one's act, and not killing chosen as a means to achieving one's end. It is therefore licit, according to Aquinas, to deliberately choose to kill an aggressor in self-defense.

Let us look more closely at Long's argument. Long notes that in Aquinas intention pertains principally to the willing of ends and not to means (Long, 2007, 41).[4] Aware that Aquinas qualifies this in several places,[5] Long explains that intention can be understood in "secondary" and "analogous" senses as extending also to means (Long, 2007, 41–42). Nevertheless, none of these should be understood as pertaining to "the *per se* instance of the human act" in which intention pertains only to last ends. When Long turns to Aquinas's article on killing in self-defense, he insists that Aquinas intends it to be read as pertaining to "the *per se* instance of human action." Since the "formal," "defining" species of the act in the per se case is derived from the end and not the moral object, and the killing (the object) is ordered to the end of self-defense, the killing "is, literally, *praeter intentionem*" (Long, 2007, 43). Long thus concludes that according to Aquinas one may licitly choose "*a deliberately and per se lethal means*" (Long, 2007, 50, emphasis in text).

Long is correct in saying that Aquinas conceives intention as pertaining principally to ends. But is he correct in insisting that Aquinas's discussion of intention in the treatise on self-defense is limited to his (Long's) "per se instance of human action"? There are four reasons concluding that it is not.

First, Aquinas does not make the strong distinction between "per se" and "secondary" instances of action that Long makes. Although Aquinas principally means intention to pertain to the willing of ends, he states in several places that ends can include mediate ends midway between one's closest-in means and one's most remote purposes for acting: "a terminus [end] is something last, not always in respect of the whole, but sometimes in respect of a part" (Aquinas, *ST*, I–II, Q. 12, a. 2, ad 2). Now if the end towards which the will inclines is a mediate end, it stands in relation to the will both as end and means. Strictly speaking it is not intended *as* means, but as (proximate) end. But precisely because it is proximate and not ultimate, it is willed also as means through which the further end is acquired. In this sense one intends one's means. For example, on a road trip from Virginia to Louisiana a man says to his children, "our goal for today is to reach Alabama." Alabama is their end, albeit a proximate end with respect to the whole trip. And it is proper to say they *intend* to reach Alabama today (as well as intend other closer-in ends: "to find a gas station," "to stop at Cracker Barrel," "to avoid a traffic ticket").

Long acknowledges this sense of "intending the means", but he excludes the possibility that it extends to Aquinas's discussion of self-defense (in Aquinas, *ST*, II–II, Q. 64, a. 7c). But he offers no *reasons* for excluding this sense. He merely asserts (again and again) that there exists in Aquinas a "*per se* case" where chosen means are distinguished from intended ends (see Long, 2007, 41, 42, 43). It is true that ends and means are analytically distinguished in Thomistic texts (for example, Aquinas, *ST*, Q. 12, a. 4, ad 3). But this fact neither supports the assertion that Aquinas means to propose a working concept of a "*per se* instance of human action," nor (a fortiori) that his treatise on self-defense is meant to be read in light of such a concept.

Because Aquinas states clearly elsewhere that intention can be both of ends and means, there is good, indeed compelling, reason for rejecting Long's interpretation. Consider Aquinas's following assertion:

> Intention regards the end as a terminus of the movement of the will. Now a terminus of movement may be taken in two ways. First, the very last terminus, when the movement comes to a stop; this is the terminus of the whole movement. Secondly, some point midway, which is the beginning of one part of the movement, and the end or terminus of the other. Thus in the movement from A to C through B, C is the last terminus, while B is a terminus, but not the last. *And intention can be both.* Consequently though intention is always of the end, it need not be always of the last end (Aquinas, *ST*, I–II, Q. 12, a. 2, emphasis added).

Aquinas plainly asserts that an object conceived precisely as a midway point, or a terminus on the way to one's last end—which in relation to that end is a *means*—can be intended. This is not an isolated example. Consider the following assertion:

> Intention is not only of the last end, as stated above, but also of an intermediary end. *Now a man intends at the same time, both the proximate and the last end* (Aquinas, *ST*, I–II, Q. 12, a. 3c, emphasis added).

In fact, in *De Malo*, Aquinas takes care to avoid the conclusion to which his account of intention inclines if read according to Long's "*per se* case" reading. Construing intention as the willing of last ends only would mean that one who wills an evil means to a good end—Aquinas uses the examples of stealing to give alms and killing the saints to render a service to God—possesses "a bad will but a good intention." Aquinas is unsettled by this, as are we, and so offers the following qualifier:

> If by intention one understands not only the intending of the end but the willing of the means [*voluntas operis*], then it is true to say that in good deeds and in evil deeds, as much as one intends, so much does one do (Aquinas, *de Malo*, Q. 2, a. 2 ad 8).[6]

In this way, if "the whole [i.e., the willing of the end and means] is designated as intention, the intention will also be evil" (Aquinas, *de Malo*, Q. 2, a. 2 ad 8). In other words, one who wills an evil means to achieve a good end is not good intentioned.

In discussing human action, Aquinas so consistently qualifies the term end to include intermediate ends,[7] that Long's claim that intention in the treatise on self-defense includes only the last end of self-defense and not the killing of aggressors as a means is rendered implausible. In Aquinas's words, "*intention can be both.*"[8]

Second, Aquinas says in question 64, article 7 that acts are morally specified by their intention, not according to what is *praeter intentionem* "since this is accidental as explained above." For Aquinas, effects that are *praeter intentionem* are "accidental." If we go to the text ("explained above") to which Aquinas refers (namely, Q. 43, a. 3), we see can see Aquinas's meaning for the term "accidental." The text treats the evil of scandal. Aquinas states that "scandal is accidental when it is beside the agent's intention (*praeter intentionem*), as when a man does not intend, by his inordinate deed or word, to occasion another's spiritual downfall, but merely to satisfy his own will" (Aquinas, *ST*, II–II, Q. 43, a. 3). The downfall of the other—the scandal—is accidental because it is *praeter intentionem*. Now Long argues that *praeter intentionem* in Q. 64, a. 7 should be read as relating only to remote ends, but not to means deliberately chosen by way of fulfilling those ends. Since Q. 43 interprets Q. 64 in regard to the term "accidental", it follows for Long, that relative to what is intended, what is willed as a means is accidental. So according to Long one may will the downfall of one's neighbor as a means to some further end, and only the end would be intended, but the means—the downfall—would be accidental because *praeter intentionem*. If, for example, a presidential candidate intent on winning an election (the *intended* end in Long's "*per se* case") does something precisely *in order to* discredit his rival by deliberately inducing him to fall into a public mortal sin (the *means* to winning), his rival's sinning would be accidental because *praeter intentionem*. The act would be, say, a mercenary act of trying to win an election, but not an *act* of scandal.

But this misses the point of Aquinas's account of scandal. Scandal is accidental—is *praeter intentionem*—Aquinas says, when one does "not intend, by his inordinate deed or word, to occasion another's spiritual downfall, but merely to satisfy his own will" (Aquinas, *ST*, II–II, q. 43, a. 3). This might be the case, for example, when a man to fulfill his lust dips into an adult bookstore to purchase pornography ("to satisfy his will") and is seen by his 16 year old son, who, waiting till his father leaves, sets out himself to satisfy desire by purchasing pornography. But this is certainly *not* the case with our presidential candidate who sets as the proximate *end* of the deliberate act of his will the downfall of his rival. He means to lead his rival into sin, and deliberately formulates and carries out a plan of action with this precise aim; but the sin is a means to winning the election. His act, according to Aquinas, and Catholic moral tradition,[9] would manifestly be an act of scandal.

It is clear then that in his treatise on scandal, both the terms "accidental" and "*praeter intentionem*" are meant to refer to effects that are willed neither as the end of one's act, nor means towards that end. To be accidental, the neighbor's downfall must fall outside the ends-means nexus as a side-effect of one's plan of action.[10] Similarly, the killing in an act of self-defense according to Aquinas is legitimate if it is neither the end nor means of one's act, but an accidental effect of the end of preserving one's life and means of proportionate force to bring that about; the killing is *praeter intentionem* and therefore accidental.

Third, Aquinas states that "the will is the principle of moral actions" (Aquinas, *ST,* I–II, Q. 20, a. 6) and that "intention, properly speaking, is an act of the will" (Aquinas, *ST,* I–II, Q. 12, a. 1). But the will stands in a threefold relation to the good (Aquinas, *ST,* I–II, Q. 12, a.1 ad 4). The first is the will's ordinary openness to and interest in human good, what Aquinas calls *voluntas simplex* or simple willing, which precedes human action; the second is the will's resting with satisfaction in the good once realized, or *fruitio*, which follows upon action; and the third is the movement towards goods proposed to the will by reason, a movement which Aquinas refers to as *intentio* (from the verb *intendere*, "to stretch towards") (Aquinas, *ST*, I–II, Q. 12, a.1). But this also defines the movement of the will in choosing, which Aquinas calls *electio*: "in so far as the movement of the will is to the means, as ordained to the end, it is called *choice* [*electio*]: but the movement of the will to the end as acquired by the means, is called *intention* [*intentio*]" (Aquinas, *ST,* I–II, Q. 12, a.1 ad 3). In both cases there is a tending of the will—a *tendere*, a "stretching towards". Choice is "that act whereby the will *tends to something* proposed to it as being good" (*voluntas tendit in aliquid*) (Aquinas, *ST,* I–II, Q. 13, a.1); *intention* is that act whereby "the will ... *tends to something* according to the order of reason" (*voluntas in aliquid tendit*)" (Aquinas, *ST*, I–II, Q. 12, a.1, ad 3). The movement of the will to the end and the movement through the means to the end are the same type of movement of the will (Aquinas, *ST*, I–II, Q. 12, a.4). Aquinas's distinction then between *intentio* and *electio* does not designate a distinction in types of acts of the will, but rather a formal distinction of the relationship of the will relative to ends and means.[11] Both entail a *tendere* towards the good as appointed by reason. If the movement is towards an end as a *terminus* of movement, Aquinas calls the movement intention (Aquinas, *ST*, I–II, Q. 12); if the movement is towards the means *for the sake of the end*, it is called choice (Aquinas, *ST*, I–II, Q. 13). But as we have said, every means between one's most proximate (close-in) means and one's last end in acting is itself an end with respect to the prior means, an end as a *terminus* of movement. The will's movement towards these proximate ends therefore entails a fully specified *intentio*.[12] Now since moral acts take their species according to what is intended (Aquinas, *ST*, II–II, Q. 64, a. 7), and the means insofar as they stand as close-in ends are intended, the means in this sense are also morally specifying. As ends in their own right, with chosen means usually required for their realization, they are a defining principle of the moral species of that act constituted by the proximate end-means movement. Insofar therefore as the will is tending towards an end, whether towards the final end alone or towards a proximate end on the way to the final end, the will's movements morally specify the action. Therefore, insofar as an individual under attack chooses to kill his assailant, the killing—and this is the crucial point Long fails to see—is *both* the means he elects to realize his end of self-preservation *and* the end relative to some chosen set of means such as shooting, stabbing, thrusting, etc. Neither proximate ends nor remote ends can be conceived as *praeter intentionem*, and so neither can deliberate killing in self-defense.[13]

Fourth, if we take Long's concept of the "*per se* instance of human action" strictly, as he wants us to, we must exclude all secondary senses of intention when interpreting question 64 article 7. We must understand Aquinas as including only the last end which moves one to act, an end which is literally the first principle of one's action: first in intention and last in execution (Aquinas, *ST*, I–II, Q. 1, a. 1, ad 1). But to read *article 7* this way is inconsistent with Aquinas's wider treatment of the lawfulness of killing (*homicidium*) in general.

The eight articles treated under question 64 (entitled *De Homicidio*) are his most developed on the subject. In articles 3, 5 and 6 he establishes respectively exceptionless negative norms against all intentional killing by private persons, the intentional killing of oneself in suicide and the intentional killing of the innocent. But the norm is not exceptionless against all intentional killing. Public authority acting on behalf of the welfare of the community legitimately kills criminals with intent (Aquinas, *ST*, I–II, Q. 1, a. 3). Aquinas writes famously in article 2 that every part is ordered to its corresponding whole as imperfect is ordered to perfect. Parts exist then naturally "for the sake of the whole." Using the example of the human body, he says that if a limb grows putrid with disease it may be useful (*expediat*) to the health of the whole body to amputate—cut off—the diseased limb. In this way a part is sacrificed for the welfare of the other members. The act of cutting in this case "will be praiseworthy and salubrious." He then applies the reasoning to the relationship between a member of the community and the community as a whole and concludes that "if any one is dangerous and corrupting to the community on account of some sin, it is praiseworthy and salubrious that he be killed, in order to preserve the common good" (Aquinas, *ST*, I–II, Q. 1, a. 2).[14]

The *homicidium* defended here is intentional. Civil authority is authorized to carry it out because it is charged with the care of the common good. A similar exception for civil authority is articulated in Aquinas's treatise on self-defense that we have been considering (a. 7). Aquinas states "it is unlawful [intentionally] to take a man's life, except for the public authority acting for the common good, as stated above" (Aquinas, *ST*, II–II, Q. 64, a 7; the internal reference is to a. 3). Civil authority lawfully intends to kill because it "refers" the killing to the public good. The principle here justifying civil authority's rightful killing is the same as it was in article 2: public authority legitimately kills with intent when the killing is referred to the common good.

But it is clear that this type of *intentional* killing (which is the normative type for justifiable intentional homicide throughout question 64) does *not* conform to Long's strict "per se instance of human action." If it did, the killing could not be said to be intended, since it is willed as a means to the end of preserving the common good. Yet Aquinas clearly calls the killing *intended*: "it is not lawful for a man to *intend* killing a man in self-defense, except for such as have public authority, who while *intending* to kill a man in self-defense, refer this to the public good" (Aquinas, *ST*, II–II, Q. 64, a 7). It therefore must be the case that Aquinas means to include under his use of intention in q. 64, a. 7 the willing of both ends and means.[15]

6.2 *Praeter Intentionem* as Including Inevitably Foreseen Side-Effects?

Another disputed question is whether Aquinas means *praeter intentionem* to include bad effects foreseen as inevitable. If an effect follows always, or almost always, from a kind of action, is not that effect so closely united to the agent's *doing* as to be inseparable from the agent's intention? There are many examples in the literature: the passenger airplane shot down to prevent terrorists using it as a guided missile; the fat man in the cave blown with dynamite out of the hole in which he is stuck to save the other members of his spelunking party whose lives are threatened by rising water; the baby stuck in her mother's birth canal threatening both her and her mother's life, whose skull is crushed by the doctor in order to remove the baby and save the mother. Can the deaths of the passengers, the fat man and the baby be *praeter intentionem* notwithstanding the fact that they were foreseen as inevitable? Joseph Boyle argues persuasively that death in cases like these need not be intended (Boyle, 1978, 658–660, 1991, 486–489, 1977). Thomas Cavanaugh disagrees.

Although Cavanaugh argues (contra Long) that *praeter intentionem* in Aquinas extends only to side-effects willed neither as the end or means of an act, he argues (contra Boyle) that in the treatise on self-defense the killing is *praeter intentionem* only when it *threatens* to follow, but not when one foresees that one's act inevitably will cause death (Cavanaugh, 1997). So Aquinas defends *risking* death in self-defense but not foreseeing it. Cavanaugh's conclusion rests on Aquinas's meaning for the term *quandoque* ("sometimes") in a reply to an objection in the article on self-defense. The objection states that since homicide is a graver sin than fornication and adultery, and it is never licit to commit fornication and adultery to preserve one's life, it is therefore never legitimate to kill another to save one's life. Aquinas replies: "the act of fornication or adultery is not ordered by necessity to the preservation of one's own life, as is the act [of self-defense] from which homicide sometimes [*quandoque*] follows."[16] Cavanaugh thinks Aquinas's use of *quandoque* demonstrates that he means to limit the scope of *praeter intentionem* to deaths which occur only sometimes (i.e., to deaths that *might* occur). Cavanaugh admits (with Boyle) that Aquinas uses the term *praeter intentionem* elsewhere to refer to consequences that inevitably or usually follow from an intended act. But he denies he means to use it this way in the treatise on self-defense: "he does not refer to an assailant's death as foreseen as inevitable" (Cavanaugh, 1997, 118). Cavanaugh thinks Boyle's failure to contend with the meaning of *quandoque* leaves his argument unpersuasive.

Cavanaugh is undoubtedly correct in saying Aquinas "does not refer to an assailant's death as foreseen as inevitable" in the treatise on self-defense. But nor does he exclude acts of self-defense in which death is foreseeable. He simply does not address directly the question of foreseeability.

Cavanaugh misunderstands Aquinas's use of *quandoque*. The statement that "homicide sometimes follows" from an act of self-defense is stating no more than that some but not all acts of self-defense bring about death. Said another way, *sometimes* an act of self-defense rightly ordered to the preservation of one's own life will *not* bring about an aggressor's death; sometimes (*quandoque*) however the aggressor's death will follow. In such a case, it is legitimate to choose such an act, provided one's purpose is to cease the aggression and not *to* kill. Aquinas's statement has no bearing on the measure of foreseeability of the aggressor's death.

Commenting on article 7, Boyle states "the effect of the assailant's being killed appears to be a foreseen and predictable consequence of the type of act of self-defense being considered" (Boyle, 1978, 658). The conclusion is presumably drawn from Aquinas's words "nor is it necessary for salvation that a man omit the act of moderate self-defense in order to avoid killing the other man." Beyond this the text implies nothing about the foreseeability of the secondary effect. To draw a strong conclusion on what Aquinas thinks about the relationship between intentionality and the bringing about of inevitably foreseen side-effects, we need to go outside the treatise on self-defense. And Boyle does this.

He begins by looking at a passage from the *Commentary on Aristotle's Physics*. Aquinas writes:

> Hence whatever takes place in the effect outside this intention (*praeter intentionem*) is *per accidens*. And I say this is true if what is outside the intention (*praeter intentionem*) follows in few cases. For what is always or frequently joined to the effect falls under the intention (*cadit sub intentione*) itself. For it is stupid to say that someone intends something but does not will (*non velit*) that which is always or frequently joined to it (Aquinas, 1999, 210).

Although on first glance this might be taken as opposing Boyle's conclusion, Boyle points out that its logic implies that what is *praeter intentionem* is not identical with what accompanies an effect rarely or accidentally. The first and second sentences are referring to effects that are *praeter intentionem*, which Aquinas calls *per accidens*. This context continues into the remaining two sentences. If what is outside the intention is always or frequently joined to the intended effect, then we should say it "falls under the intention" (i.e., is joined to the intention), which is different from saying it is intended. Aquinas says it cannot be said not to be *willed*. Thus, Boyle argues, "what falls under the intention and is therefore inseparable from what is intended can be *praeter intentionem*" (Boyle, 1978, 659).

Boyle supports his reading with a text from Aquinas's *De Malo*.[17] Aquinas speaks of cases where a *per accidens* effect accompanies an intended effect "always or in most cases." In such an instance, "the accidental effect is not separated from the agent's intention." As in the *Commentary on Aristotle's Physics*, Aquinas identifies the *per accidens* effect with the agent's intention, yet still says it is accidental. He says if some evil always accompanies an intended good, the actor is "not excuse(d) from sin, even if that evil is not *per se* intended." So he obviously believes that effects inevitably foreseen can be brought about *praeter intentionem*. But they are in some way inseparable from what is intended. Boyle suggests they be called "intended *per accidens*" (as opposed to "*per se* intended") (Boyle, 1978, 660).

Boyle's term does not seem entirely adequate since an effect is *per accidens* precisely in light of it falling outside the intention, which obscurely implies for Boyle's term that the *per accidens* effect is intended *praeter intentionem*. Yet Boyle is making an important point. Since the inevitably foreseen side-effect cannot be separated from the agent's intention, it does not seem right to refer to it as *un*-intentional. It is certainly brought about willingly, since the agent could have chosen not to act. This makes understandable the conclusion of those who say it is necessarily intended. But Aquinas does not say that what is always or frequently joined to an intended effect is intended; he says rather it is foolish to say one does not *will* (*non velit*) it. Perhaps a better term is to call it a "voluntary effect."[18] We can then say that the inevitably foreseen side-effect is voluntarily brought about, though not *per se* intended.[19]

Aquinas obviously thought the distinction between intentional effects and voluntary effects brought about *praeter intentionem* was important enough to repeat several times.[20] Since he clearly denies that foreseeable effects are necessarily intended, he cannot reasonably be read as holding that they are necessarily intended in the context of self-defense. Boyle is therefore correct in concluding strongly that Aquinas should be read as asserting that though an aggressor's death can be foreseen with certainty to follow from one's act of self-defense, it is not *per se* intended (Boyle, 1978, 660). Whether foreseeable effects are intended depends upon the structure of intending, not on the content of the effects.

Steven Long follows Cavanaugh in relying on the content of effects to criticize Boyle's view. Boyle argues that "an action carries out a proposal adopted by choice" and that "one intends both to carry out that proposal [i.e., intends the means] and to achieve the purpose for the sake of which one adopts it [i.e., intends the end]" (Finnis et al., 2001, 7). The proposal is an intelligible plan one believes is apt to bring about some desired state of affairs.

Long criticizes this view as unreasonably formalistic, "approximating," in his words, "the thought of G.E. Moore" (Long, 2007, xi, note 1). His conclusion stems from his understanding of the "object of the act." Aquinas, Long says rightly, defines the object as something "relative to reason" (Long, 2007, 11). But this relation to reason, Long holds, "always materially includes the act itself and its integral nature" (Long, 2007, 11). The term "the act itself and its integral nature" is original to Long and does considerable work in his argument. It implies that we need to factor into the definition of the moral object a certain "material aspect", what Long refers to as an act's "physical structure" (Long, 2007, 12, 108). This might include physical effects causally consequent upon a piece of chosen behavior, spatial proximity of the actor to the effect, foreseeability that one's behavior will cause an effect, etc. Long argues that this "material aspect" rules out the possibility that the moral object is an exclusively formal dimension of an act, even though Aquinas likens the object to a thing's form (Aquinas, *ST*, I–II, q. 18, a. 2). The object is only formal insofar as it is morally defining, but that by which it is so defining necessarily includes a material dimension: "the object of the act is formal in the sense of most determining and defining, but . . . *within* this object there is also a relative distinction in terms of a formal and a material part" (Long, 2007, 13).

Where in Aquinas does Long find this necessary "material" aspect? He appears to derive it from Aquinas's discussion of "external" actions (Aquinas, *ST*, I–II, Q. 20). In explaining human acts, Aquinas distinguishes between the "interior act of the will" (or the will's "inner object", i.e., the good as apprehended and proposed as choiceworthy by reason and as intended by will),[21] and the "external action" by which the choice for the object is carried out in our behavior. The "inner object" and "external action," Aquinas says, are related as form and matter, as the *materia circa quam* and the *materia ex qua* of an act.[22]

Aquinas identifies two fonts of goodness and badness in external acts, one proceeding from the orientation of the will and the other depending on the contents of reason. What concerns the "order to the end" in an act, he says, depends entirely on the will, since it is the will that moves towards ends. What concerns the "due matter (*debita materia*) and circumstances" of the act, however, depend on reason since reason envisages possibilities for action and proposes them as ends to be pursued (Aquinas, *ST*, I–II, Q. 20, a. 2). The due matter is what reason envisages and proposes to will as a way of bringing about some purpose. It is the intelligible content of a practical "what" or proposal to *do* something. Aquinas says that on this depends the goodness of the will (Aquinas, *ST*, I–II, Q. 20, a. 2). This is also how Aquinas defines the moral object: constituted by reason's proposal and determining the goodness of the will. (Aquinas, *ST*, I–II, Q. 19, aa. 1–3).

It is in Aquinas's term *debita materia* that Long finds the concept of a necessarily material dimension for the moral object. He says the due matter "is materially included in the object of the act, for it makes up part of the very definition of the act, and as such must be materially presupposed in the object as "what *the act* is about relative to reason" (Long, 2007, 12–13, note 13). If this meant merely that the intelligible practical possibility (the due matter) has a determinate relationship to envisaged states of affairs in the world, it would be unproblematic. But Long means more than this. He means that the contents of the moral object are limited by external physical states of affairs even when those affairs are not part of one's practical interests. *Debita materia* however refers simply to a practical possibility that one believes can be achieved through action, one deems choiceworthy and one formulates as a plan of action. Most practical plans are envisaged as having an external manifestation in the physical world; and to the extent that one's plan includes this external manifestation (e.g., smashing *this* window, killing *this* unborn baby), its bringing about is part of what one chooses. As part of reason's proposal, it constitutes part of the moral object (Aquinas, *ST*, I–II, Q. 19, aa. 1–3).

But it is important to see that the contents of one's plan and hence one's moral object are determined by what one is interested in. Long thinks that external physical conditions other than what one includes in one's intelligible plan are necessarily determining of the moral object (e.g., physical conditions causally related to one's intelligible proposal, spatial proximity of the body of the actor to some effect, foreseeability of side-effects one is *un*interested in bringing about). And so he inserts his own condition that "the act itself and its integral nature are always materially included in the object" (Long, 2007, 12). Long wants to assure that the content of the moral object does not get cut off from determinate conditions in the real world, lest an act becomes no more than "a placeholder standing for we know not what" (Long, 2007, 13). But in his attempt to ground his account in what he believes to be a more adequate epistemology, Long turns away from Aquinas. Neither the *per accidens* effects of an act, nor external descriptors identifiable by third parties are what one looks to in defining the moral object, but rather what one is interested in doing (i.e., one's intelligible plan for bringing something about). Aquinas's concept of due matter refers to *what* one is interested in bringing about. The "what" includes certain states of affairs judged choiceworthy to bring about either for themselves or for what they make possible. This does not include the causally related physical dimensions that Long says are necessarily included in the object. Therefore his accusation that those who deny a necessary physical dimension to the moral object unwittingly espouse a "pure *angelism*, a residue of Cartesian error" (Long, 2007, 108) is either unfair, or must extend likewise to Aquinas.

In effect Long is arguing that if an agent causes physical effect X, knows in advance he will cause X, and X follows with immediate physical proximity to the agent's intended behavior (follows "directly" in a *physical* sense), then effect X is necessarily part of the moral object intended by the agent.

For example, Long addresses the use of methotrexate in treating an ectopic pregnancy. He judges the use of methotrexate to be an evil means because "its natural effect is the destruction of tissue essential to the life of the *conceptus*" (Long, 2008, 97). Methotrexate's mode of action is to interfere with the synthesis of DNA in the trophoblastic tissue thus causing the implanted embryo to be expelled from the wall of the fallopian tube. Long refers to the act by which one administers methotrexate to treat an ectopic pregnancy as "terminating in the . . . poisoning" of the conceptus. His term "terminating in poisoning" is telling. It refers to the causal consequences of using methotrexate. And indeed the poisoning of the conceptus is a causal consequence. But Long thinks that because death results from its use and that because one foresees that death will result, choosing to use methotrexate is not simply tolerating a foreseen side-effect in the treatment of ectopic pregnancy, but is necessarily choosing to kill the conceptus as the means to assisting the woman. Because the physical effect follows with causal predictability, Long thinks the act's moral object necessarily includes the killing. He does not believe the choice can be separated from the effect brought about by the choice.

Likewise, he judges that removing a conceptus to treat ectopic pregnancy through the microsurgery known as salpingostomy is intrinsically evil since it is an act "terminating" in the "spearing . . . of the conceptus" (Long, 2007, 97). In salpingostomy a slit is made in the fallopian tube at the site of the embryo's nesting; the embryo is scraped off the wall of the tube or plucked off with forceps along with some damaged tissue; this usually kills the embryo.[23] Since the physical behavior is carried out on the conceptus, and the behavior results in death, the moral object must include the killing of the conceptus. Again, Long focuses on qualities of the physical effects—their causal consequence, their proximity to the actor's behavior—to define the moral object. He fails to follow Aquinas in assessing the moral object in terms of the *reasons* moving the agent to act

Long approves however the procedure called salpingectomy to treat ectopic pregnancies. A damaged fallopian tube is cut below and above the site of the misplaced conceptus and the piece of damaged tube along with the conceptus is removed. The conceptus, of course, inevitably dies, but because "this motion does not terminate in *death*" it can be morally legitimate (Long, 2007, 97). The external behavior—the "motion"—and not reason's proposal is decisive. Again, Long sticks to the principle that "the physical structure of the act materially enters into the moral object" (Long, 2007, 108).

Boyle's account would argue that neither the use of methotrexate nor salpingostomy need entail the intent to kill either as a proximate or remote end of the act.

In both, the intention of a morally upright person would be to carry out a proposal to remove an implanted conceptus from an anomalous and dangerous location in a woman's fallopian tube. This proximate end (or *means*) is sought in order to save a woman from death or grave harm (the remote *end*). The due matter then is the *removal of the conceptus*. The death follows inevitably from the removal; it is a foreseen side-effect. It is a side-effect precisely because it is not part of the proposal adopted to remove the conceptus for the sake of the physical integrity of the woman (Finnis et al., 2001, 3). It is what we referred to above as an effect voluntarily brought about but not *per se* intended. The killing is *praeter intentionem*.

Notes

1. For a more in-depth consideration of the concept of intention and related concepts in Aquinas and John M. Finnis, see Brugger (2005).
2. Boyle's defense of this position is found at Boyle (1978, 657–660); for "intention" as related to the chosen proposal (plan of action) formulated for realizing a purpose, see Finnis (1986, 1996). Cf. Boyle (1978, 664).
3. Long is particularly critical of the interpretation proposed by Germain Grisez, John Finnis and Joseph Boyle (sometimes called the New Natural Law Theory). Although he states that his text is not meant to be a polemic (p. xii), much of it is an extended polemic against Grisez, Finnis and Boyle (see Long's comments on pages xi–xv, 39–40). For the New Natural Lawyers' account of action theory, see Finnis et al. (2001).
4. Long references Aquinas (*ST*, I–II, Q. 12, a. 4, ad 3).
5. See for example Aquinas (*ST*, I–II, Q. 12, a. 2; ad 2; a. 3c; *SCG*, Bk. 3, ch. 2, no. 5; see also *De Malo*, q. 2, a. 2 ad 8).
6. See also *In Sent* (II, d. 36, q. 1, a. 5, ad 5; *In Sent. II*, d. 40, q. 1 a. 2c). For Aquinas on "object" as (proximate) end, see *In Sent* (II, d. 36, q. 1, a. 5, ad 4; *ST*, I–II, Q. 72, a. 3, ad 2).
7. "For things which act for an end, all things intermediate between the first agent and the ultimate end are as ends in regard to things prior, and as active principles [means] with regard to things consequent" (Aquinas, *SCG*, Bk. 3, Chapter 2, no. 5); see also Aquinas, *Physics*, Bk. 2, lect. 5, n. 6 [181].

8. Finnis, Grisez and Boyle write: "The reason why both the broader and the narrower senses of 'intention' are appropriate is that the distinction between proximate and further objectives is highly relative. Every means one adopts in the pursuit of some end will also be an end whenever there is a prior means—one closer in to the agent" (Finnis et al., 17).
9. "Scandal is a grave offense if by deed or omission another is *deliberately* led into a grave offense" (*Catechism*, 2284, emphasis added).
10. This point is well illustrated by *Case A* (of the hungry boy) in Finnis et al. (2001, 3).
11. Aquinas is not saying that the objects are the same in choice and intention; in fact they are different (see *ST*, I–II, Q. 12, a. 4, ad 2). But the types of movement are the same (*ST*, I–II, Q. 12, a. 4, ad 3).
12. Philippa Foot says something similar: "[one] intends in the strictest sense both those things that he aims at as ends and those that he aims at as means to his ends" (Foot, 2001, 144).
13. This is why John Paul II solemnly condemning abortion in *Evangelium vitae* adopted the formulation: "I declare that direct abortion, that is, abortion as an intended end or as a means ["*uti finem intentum seu ut instrumentum*"], always constitutes a grave moral disorder, since it is the deliberate killing of an innocent human being" (John Paul II, 1995, no. 62). See also Paul VI: "Similarly excluded is any action which either before, at the moment of, or after sexual intercourse, is specifically intended to prevent procreation—whether as an end or as a means" (Paul VI, 1968, no. 14).
14. I criticize Aquinas's account of legitimate intentional killing in capital punishment in Brugger (2003, 169–176); and Brugger (2004, 357–372). In Brugger (2008, 392–394), I utilize Aquinas's account as a springboard for a development of doctrine in Catholic teaching that would reject in principle capital punishment as retributive killing.
15. Moreover, in reply to objection 1, Aquinas, commenting on Augustine, states that it is not licit "when one man intends to kill another to save himself from death" (Aquinas, *ST*, II–II, Q. 64, a 7, ad 1). Killing here is clearly the means since the end is specified as "to save himself from death."
16. "[A]ctus fornicationis vel adulterii non ordinatur ad conservationem propriae vitae ex necessitate, sicut actus ex quo quandoque sequitur homicidium" (Aquinas, *ST*, II–II, Q. 64, a. 7, ad 4).
17. Aquinas (*De Malo*, Q. 1, a. 3, ad 15): "Ad decimumquintum dicendum, quod aliquando accidens alicuius effectus coniungitur ei ut in paucioribus et raro; et tunc agens dum intendit effectum per se non oportet quod aliquo modo intendat effectum per accidens. Aliquando vero huiusmodi accidens concomitatur effectum principaliter intentum semper, vel ut in pluribus; et tunc accidens non separatur ab intentione agentis. Si ergo bono quod voluntas intendit, adiungitur aliquod malum ut in paucioribus, potest excusari a peccato; sicut si aliquis incidens lignum in silva per quam raro transit homo proiiciens lignum interficiat hominem. Sed si semper vel ut in pluribus adiungatur malum bono quod per se intendit, non excusatur a peccato, licet illud malum non per se intendat."

18. See my discussion of the relevance of this distinction for understanding the self-determining effects of action in Brugger (2005, 90–92).
19. Anscombe uses a similar expression: "voluntary though not intentional" (Anscombe, 1957, 89); see also Brock (1998, 97). Hart suggests using the term "oblique intention" (distinguished from "direct intention") to indicate that an agent does not bring them about "*un*intentionally" (Hart, 1968, 120–121); see also Williams (1965, 10). Finnis agrees we should avoid saying they are caused unintentionally, since unintentional implies "accident or mistake or lack of foresight." But he rejects the term "oblique intention" (Finnis, 1991, 47–48).
20. In another place Aquinas writes that "not everything that is *praeter intentionem* is necessarily fortuitous or a matter of chance. . . . For, if that which is *praeter intentionem* be a consequence that follows either always or frequently of what is intended, then it does not occur fortuitously or by chance" (Aquinas, *SCG*, Bk. III, Chapter 6).
21. On the interior act of the will, see *ST* (I–II, Q. 19; 18, a. 2c, ad 2–3).
22. On this point, see Gleeson (1999, 361).
23. Christopher Kaczor refers to a documented case where an embryo after salpingostomy was able to be resettled in the uterus and resulted in a healthy birth (Kaczor, 1998, 356).

References

Anscombe, G.E.M. 1957. *Intention*. Oxford: Basil Blackwell.

Aquinas, T. 1999. *In Octo Libros Physicorum Aristotelis Expositio, Commentary on Aristotle's Physics*. (Eng. trans: Blackwell, R., Spath, R., and Thirlkel, W.E. Notre Dame, IN: Dumb Ox Books.

Aquinas, T. *Quaestiones Disputatae de Malo (de Malo).*

Aquinas, T. *Scriptum super Libros Sentiarum (In Sent.)* I and II.

Aquinas, T. *Summa Theologiae (ST).*

Aquinas, T. *Summa Contra Gentiles (SCG).*

Boyle, J.M. Jr. 1977. Double-effect and a certain type of embryotomy. *Irish Theological Quarterly* 44:303–318.

Boyle, J.M. Jr. 1978. *Praeter Intentionem* In Aquinas, *The Thomist* 42:649–665.

Boyle, J.M. Jr. 1991. Who is entitled to double effect? *The Journal of Medicine and Philosophy* 16:475–494.

Brock, S. 1998. *Action and conduct: Thomas Aquinas and the theory of action*. Edinburgh: T & T Clark.

Brugger, E.C. 2003. *Capital punishment and Roman Catholic moral tradition*. Notre Dame, IN: University of Notre Dame Press.

Brugger, E. C. 2004. Aquinas and capital punishment: The plausibility of the traditional argument. *Notre Dame Journal of Law, Ethics & Public Policy* 18:357–372.

Brugger, E.C. 2005. Action, intention and self-determination. *Vera Lex: Journal of the International Natural Law Society*, New Series 6:79–106.

Brugger, E.C. 2008. Rejecting the death penalty: Continuity and change in the tradition. *The Heythrop Journal* 49:388–404.

(1997). *Catechism of the Catholic Church: With Modifications from the Editio Typica, 2nd edition*. Washington, DC: U.S. Catholic Conference.

Cavanaugh, T.A. 1997. Aquinas's account of double effect. *The Thomist* 61:107–121.

Finnis, J.M. 1986. The act of a person. *Persona Verità e Morale: Atti del Congresso Internazionale di Teologia Morale.* Roma: Città Nuova Editrice, 159–175.

Finnis, J.M. 1991. Intention and side-effects. In *Liability and responsibility: Essays in law and morals*, eds. R.G. Frey and C. W. Morris, 32–64. Cambridge, MA: Cambridge University Press.

Finnis, J.M. 1996. Intention in tort law. In *Philosophical foundations of tort law*, ed. D. G. Owen, 229–247. Oxford: Clarendon Press.

Finnis, J., Grisez, G., and J. Boyle. 2001. 'Direct' and 'Indirect': A reply critics of our action theory. *The Thomist 65*:1–44.

Foot, P. 2001. The problem of abortion and the doctrine of the double effect. In *The doctrine of double effect: Philosophers debate a controversial moral principle*, ed. P.A. Woodward, 143–155. South Bend, IN: University of Notre Dame Press.

Gleeson, G. 1999. 'Medical Management' of ectopic pregnancy. In *Issues for a Catholic bioethic*, ed. L. Gormally, 353–370. London: The Linacre Centre.

Hart, H.L.A. 1968. Intention in punishment. In *Punishment and responsibility: Essays in the philosophy of law*, 113–135. Oxford: Clarendon Press.

John Paul II. 1995. *Evangelium vitae: Encyclical on the value and inviolability of human life.* Vatican City: Libreria Editrice Vaticana.

Kaczor, C. 1998. Is the 'Medical Management' of ectopic pregnancy by the administration of methotrexate morally acceptable? In *Issues for a Catholic bioethics: Proceedings of the international conference to celebrate the twentieth anniversary of the foundation of the Linacre centre 28–31 July 1997*, ed. L. Gormally, 353–370. London: The Linacre Centre.

Long, S.A. 2007. *The teleological grammar of the moral act.* Naples, FL: Sapientia Press.

Paul VI. 1968. *Humanae vitae.* Vatican City: Libreria Editrice Vaticana.

Williams, G. 1965. *The mental element in crime.* The Hebrew University of Jerusalem, Lionel Cohen lectures. Jerusalem & Oxford: Magnes Press.

Chapter 7
The Action-Omission and Double Effect Distinctions

Timothy Chappell

One crucial question for the Natural Law Theory of ethics that has been advocated by Joseph Boyle over the course of his distinguished career is: Exactly what is to count as (minimal) *respect* for a good or value? Under what circumstances do we fail to reach the threshold even of respect for some good, and so count as *violating* it? I take this to be the hardest question of all for the New Natural Law Theory; it is obvious from Boyle's writings that he takes it to be a crucial question too, and that he has done more than most to clarify it.

Whatever the Natural Law Theorist says about the boundaries of respect and violation, those boundaries are very likely to be set, at least in part, by the ways in which we articulate two famous and venerable distinctions about action: the action-omission and double-effect distinctions (henceforth AOD and PDE). My tribute here to Joseph Boyle is an examination of these two distinctions.

7.1 The Two Distinctions, and How to Argue for Them

As I understand them, AOD and PDE are the following theses:

> The action/omission distinction: *Ceteris paribus*, we are less responsible or culpable (and possibly not responsible or culpable at all) for our omissions than for our actions.

> The principle of double effect: *Ceteris paribus*, we are less responsible or culpable (and possibly not responsible or culpable at all) for our actions under those descriptions under which we do them knowingly but do not intend them, than we are for our actions under those descriptions under which we do them knowingly and do intend them.

The simple strategy of argument for these two theses goes as follows:

1. There are (at least rough) degrees of actionhood.[1]
2. Degrees of actionhood depend on proximity to a range of paradigm actions: that is, a given performance's degree of actionhood is directly proportional to

T. Chappell (✉)
Department of Philosophy, Ethics Center, The Open University, Milton Keynes, MK7 6BJ, UK
e-mail: T.Chappell@open.ac.uk

C. Tollefsen (ed.), *Bioethics with Liberty and Justice*, Philosophy and Medicine 110,
DOI 10.1007/978-90-481-9791-0_7, © Springer Science+Business Media B.V. 2011

its degree of similarity to those paradigms. This similarity can be measured by the degrees to which a given performance satisfies the conditions of actionhood.
3. There are at least three conditions of actionhood. In a paradigm action, what's done (a) could have been otherwise, (b) intentionally enacts a plan which specifies a particular set of means to a particular end or ends (or specifies the bringing-to-be of a particular instantiation of a particular value), leaving other things either (i) to happen as they foreseeably will or (ii) unconsidered and out of account, and (c) involves a causal intervention in the world. There may be other conditions of actionhood but these, so far as I can see, are the three that count.
4. *Ceteris paribus*, degrees of responsibility[2] and culpability track degrees of actionhood.
5. *Ceteris paribus*, omissions involve a lower degree of causal intervention in the world than actions do.
6. Therefore (cp. 3(c)), *ceteris paribus*, omissions have a lower degree of actionhood than actions.
7. *Ceteris paribus*, what is foreseen but not part of an agent's plan either as end or as means is less intended than what is part of the agent's plan either as end or as means, being only (i) foreseen or (ii) unconsidered.
8. Therefore (cp. 3(b)), *ceteris paribus*, actions have a lower degree of actionhood under (i) unintended but foreseen descriptions than they do under intended descriptions, and a lower degree still under (ii) unintended and unforeseen descriptions.
9. Therefore (cp. 4), *ceteris paribus*, we are less responsible and/ or culpable for our omissions than for our actions (which vindicates AOD), and for our actions under unintended descriptions, foreseen or unforeseen, than under intended descriptions (which vindicates PDE).

In Sections 7.2, 7.3, and 7.4 I shall defend this argument step by step.

7.2 Spelling Out the Simple Strategy: (1–4)

1. There are degrees of actionhood. This is a platitude; we judge agents' performances as more or less actions of theirs all the time. It's equally a platitude—and it's convenient to take the two platitudes together—that *(4) degrees of responsibility and culpability track degrees of actionhood* (*ceteris paribus*). What I mean by saying that I couldn't help pouring my coffee all over you, because the bang of the firework made me jump, is that my spilling of the coffee was *less of an action* than it perhaps seemed to you, and *therefore*, something for which I was *less responsible* than you may have thought. What you mean by saying that you acted out of character when you sexually harassed the waiter in the wine bar is that this was a slip of some sort, not an action that "the real you" caused in quite the normal way. Hence it was not fully your action; hence it was not something for which you are fully responsible. Again, the point of saying that you regret treading on my toe to get out of the way of the runaway train is that you weren't *aiming* to tread on my toe, simply to avoid the train. Though *tread on my toe* and *evade the train* were both *things that you did*, it makes sense to say that treading on my toe was less truly your action than evading the train was. What a defendant in court means by pleading "diminished responsibility" for a crime is (not, usually, that he bore *no* responsibility for that crime, but) that he was *less responsible than usual*. Such a plea is meaningless unless it is backed up by evidence that the defendant, at the time and in respect of the crime, was (not, usually, no agent at all, but) *less of an agent than usual*. In politics, law, and private life, the assumptions that agency generally explains responsibility, and that neither agency nor responsibility is all-or-nothing (although of course there are unequivocal cases of action and non-action, of responsibility and non-responsibility), are everywhere in our ordinary thinking about agency and responsibility.

Alongside this appeal to common sense and its platitudes, another way of supporting (1) is by arguing for (2) and (3). If there are, as (3) says, conditions of actionhood which articulate (2)'s notion of closeness to a paradigm action, then this will be evidence that (1) actionhood comes in degrees, both because a given performance could, presumably, satisfy one of the tests but not all of them, and also because there are degrees of satisfaction for each of the tests. If we are looking for spectrums of causation or intention or ability to do otherwise, we not only find some; we find a bewildering variety of such spectrums. But if there are degrees of satisfaction of the three tests for actionhood, there must also be degrees of actionhood itself; similarly, as (4) points out, there must also be degrees of responsibility. And so it turns out when we turn to (2) and (3). The study of excuses, as J.L. Austin noticed long ago (Austin, 1956), is a rich and varied study.

Steps 2 and 3: the conditions of actionhood. (3) spells out what (2) says, so it is (3) that we need to focus on. (3)'s condition (c) of actionhood says that a paradigm action involves a causal intervention in the world; that seems like another platitude, so I will not dwell on it.[3] Of (3)'s other two conditions of actionhood, (a) mentions "ability to do otherwise." In the sense that I mean here, this ability should not be thought of as at issue between determinists and libertarians. Sensible determinists allow that I can generally do what I choose to do, and that my choosing to do it is crucial to the causal story behind what I end up doing. The difference between determinists and libertarians is not over this, but over the conditions governing what I can choose: see Frankfurt (1969) and the vast literature that Frankfurt's paper has generated. All (a) says is that what counts as my action depends on what (else) I can choose or could have chosen at the time when I act. If I am borne helplessly down the railway line on a runaway trolley, then my ride on the trolley is not something I *do*, as it is if I climb aboard and push the trolley off at the start of the run; it's something that *happens* to me. By contrast, which levers I pull or leave alone while on the trolley *is* within the category of things that I do. The notion of what I *do* do is then obviously relative to the notion of what I *can* do.

The account of paradigm action that (3)'s condition (b) suggests may seem controversial or even question-begging in its rejection of what we might call the "global" view of agency—the view on which every action is to be seen as a contribution to the way the world is as a whole, which is perhaps the view that underlies all or at least many objections to AOD and PDE. The strength of (b) is its match with the reality of the way actual agents really behave, and with the only way they could conceivably behave. Any real agent's choice will nearly always involve his formulating a plan of action with a very limited scope, with the great majority of what exists simply lying outside that scope. That the agent must focus his intentions on the things mentioned in his plan already implies, unless that plan mentions *everything,* that there will be other things that he does not focus his intentions on, either (i) because they are not central to his plan even though he foresees them, or (ii) because he does not even foresee them. That there will be some things a finite agent does not foresee can hardly be controversial. And that he must leave other things to happen as they will, foreseeing these outcomes but not intending them—which *is* controversial—follows from the local conception of intentional action.

It also follows, interestingly enough, from value pluralism, the intuitively appealing view that there is a variety of basically different goods or values in the world, not just one form of value. The global view of agency and monism about the good are made for each other. If *per impossibile* my agency could be such that it quite generally and typically addressed the world as a whole, that would have to be because my agency could simply improve the world as a whole, without this prospect raising any insoluble questions about how to compare one aspect, in which the net value score has gone up, with another aspect in which it has gone down. (Does agency address the world as a whole wherever it finds a Pareto-optimal solution? No, even then it leaves many things entirely out of account; and in any case Pareto-optimal solutions are far from ubiquitous.) What happens in typical agency is that I aim to improve things in some particular respect, by way of some particular means, and *with no regard* to things outside my plan. There is no such thing as the single overall good, but rather lots of different separate goods; that is why there will always be indefinitely many *other* respects which don't come into my means-end plan at all. About some of these respects which lie outside my means-end plan, it will be possible and reasonable for me to ask whether my means-end plan imposes acceptable costs on them; but it won't be possible to bring them *into* that plan. About other value-aspects again my plan will simply be silent: it won't even treat them under the heading of foreseen but not intended side-effects.

4. Ceteris paribus, degrees of responsibility track degrees of actionhood. I have already noted the near-platitudinous nature of this claim. If someone wanted to find a counter-example to it, they might try negligence. In negligence, it could be argued, I am responsible precisely for *failing* to exercise actionhood; so degrees of responsibility can't always track degrees of actionhood. This objection fails because (4) does not say that the tracking relation between degree of responsibility and actionhood always holds; it says that it holds *ceteris paribus*. It will hold *ceteris paribus* if there is the following asymmetry: it takes a special reason to make me responsible for my non-action, and it takes a special reason to make me non-responsible for my action. But according to near-universal intuitions, there is this asymmetry, and negligence is part of it. "Negligence" is the name for non-action where there was a special reason why a person should have acted, and therefore is responsible for failing to act. So not only is negligence not a counter-example to (4). It is actually part of the pattern that (4) describes.

The remaining five steps of the argument are the claims (5) that omissions and (7) that what is foreseen-but-not-intended are deficient in respect of two of the tests of actionhood, and therefore (6, 8) in degree of actionhood, and therefore (9) in responsibility—QED. I will separate out the steps of this argument that relate to AOD from those that relate to PDE, focusing on AOD in Section 7.3 and PDE in Section 7.4.

7.3 Spelling Out the Simple Strategy: (5), (6), (9), and AOD

(5) says that omissions have a lower degree of actionhood than actions. This would be disputed, of course, by anyone who accepted the Symmetry Thesis—the view that all behaviour consists in or supervenes on manipulation of the body, and that keeping the body still is just as much manipulating it as moving it is. On this view, actions and omissions alike are simply forms of manipulation; and manipulation is what we are responsible for; so there is no reason to see us as less responsible for our omissions than for our actions.

Maybe something like the Symmetry Thesis is behind Bennett's view that an agent's relevance to an upshot is positive if most of the ways she could have behaved at the time would not have led to the upshot, and otherwise negative (Bennett, 1998). Bennett takes this distinction to be the nearest thing that makes sense to the traditional AOD. But does Bennett's own distinction make sense? I doubt it, because we have no idea how to individuate "ways someone could have acted," and Bennett does not tell us. He offers us a metaphor of "dividing an action space," but this is not an explanation. We cannot locate "ways of acting" in that space unless we already know how to individuate them. Without an account of how non-arbitrarily to individuate ways of acting, we cannot give sense to Bennett's phrase "most of the ways." Unless we can give sense to that, Bennett's distinction has no determinate meaning.

In any case the Symmetry Thesis is clearly false. It just isn't true that all behaviour consists in or supervenes on manipulation of the body, and that keeping the body still is just as much manipulating it as moving it is. This thesis fails to take account of the asymmetry between movement and non-movement that arises from inertia. The default condition of my body is (normally at least) not an equipoise between movement and non-movement; it is non-movement. (That is, non-movement relative to my inertial frame: it is not easier to stay in the same place relative to the sun than in the same place relative to the earth's rotating surface.) As we all know to our cost, it takes effort and decision to get things done; much less effort to fail to do them. Since the non-movement of our bodies is asymmetrically related in this way to their movement, there is a parallel asymmetry regarding our responsibility for the actions that consist in or supervene on[4] their movement and non-movement. This asymmetry takes the form already described in Sect. 7.2 discussion of negligence: we are not responsible for how we don't move our bodies unless there is a special reason why we should be; we are responsible for how we do move our bodies unless there is a special reason why we shouldn't be. This intuitive asymmetry puts paid to the Symmetry Thesis.

What about "negative actions," as they are usually called—cases where I act if and only if I don't move? There are at least a few cases that seem to deserve the description. To cite the commonest example, it is at least alleged—I've no idea whether it's actually true—that some people bid at auctions by not moving. A less convention-dependent example would be this: resolving to kill the Swiss president by ski-jumping onto his head, I make the movement that launches me down the slope of the ski-jump on the correct trajectory to perform my heinous crime, *and then keep still until I've hit him.* A third example is Nicholas Denyer's (1997). Crossing the deck of a busy diving-boat, I inadvertently stop so that I am standing on the air-pipe that supplies oxygen to one of the divers. When I realise that I am blocking his oxygen supply, I deliberately stay put. For a fourth example, take the wrestler who stands fast: his not moving from his position despite his opponent's shoving and tugging certainly looks like an action of non-movement.

Don't stories like these tend to support the Symmetry Thesis' claim that that keeping the body still is just as much a case of actionhood, and so of responsibility, as moving it is? No, they do not. I say that the Symmetry Thesis is false because there is an asymmetry between action and non-action which arises from inertia: it is normally the default condition to do nothing rather than to act, to remain at rest rather than to move. But note the word "normally:" *sometimes* movement is the default condition. This is clearest, perhaps, with the wrestler, though the ski-jumper may find it hard to keep his body in the right shape too,[5] and the diver-murderer, if he has scruples, may find it morally hard to stay put even though no physical effort is involved.[6] (As for the auction-bidder, his non-movement qualifies as an action in virtue of a very special semantic convention—one that could not work quite generally, and which is not entirely easy to imagine in full detail even in the very particular circumstances of an auction: what *particular types* of immobility should the auctioneer take as cues?) In the non-standard cases where inertia means movement, action will be resistance to that movement; negative actions will be examples of such resistance. My point against the Symmetry Thesis (my *a*symmetry thesis) is not that there cannot be cases where the default condition is movement rather than rest, but rather that such cases are non-standard.

If the Symmetry Thesis is false, then it seems obvious that omissions at least have a *lower* degree of actionhood than actions. I want to argue for something stronger: that omissions, at least in respect of the causation test, have a *zero* degree of actionhood.[7] They may have more than zero in the other two respects. In particular, an omission can obviously be intentional, and part of a plan. And when an omission is intentional, it will naturally be something that could have been done otherwise. In these ways omissions can have some degree of actionhood; but they can never have the slightest degree of actionhood by being causal interventions.

My reason for saying that omissions, at least in respect of the causation test, have a zero degree of actionhood comes from my theory of causation. Causation can be analysed lots of ways. My model is a *transmission of energy* model: "A causes B to be F" means that A transmits[8] to B the energy in virtue of which B is or becomes F. A causes B in object or context C when (and only when) A transmits to C the energy that is manifested in B's occurring. (So, e.g., the red ball causes the black ball's movement when the red ball transmits to the black ball the energy that is expressed by the black ball's movement).[9] On this model it's straightforward and obvious that omissions are not causes. What transmits energy is the cause; omissions never transmit any energy; so omissions are never causes.

This transmission or transference view of causation has been, until recently, a surprising absentee from the normal menus of alternatives in the theory of causation. Take, for instance, Jaegwon Kim's excellent article on "Causation" in *The Cambridge Dictionary of Philosophy* (Kim, 1995), which canvasses the alternative merits and difficulties of four views of causation: the counterfactual, the regularity, the manipulation, and probability views. Of these, the *manipulation* view is radically false. It says that for A to be a cause of B is for it to be true that we could, in principle, use A to produce B, or not-A to produce not-B. This gets the order of explanation between causation in general and specifically human manipulation entirely back to front. The *regularity* view is either false or vacuous. It says that there can be no instance of causation unless that instance is a token of some law-like regularity. Like certain forms of the universalisability thesis, this is false if we are stringent about what we will count as a lawlike regularity but vacuous if we are lax. In any case it could not, on its own, give us a full account of what causation is.[10] The *probability* theory states what is at most a necessary condition of a causal relation, viz., that the cause should raise the probability of its effect.[11] The faults of the *counterfactual* analysis deserve a longer examination, which they will get in a minute. The transmission view of causation is none of these four as listed by Kim. Thanks to the work of Wesley Salmon (1998) and Phil Dowe (2000, 2004), it is at least now on the map.[12] This is the account of causation that I would wish to defend (though obviously, I cannot defend it in full in the present context).

Consider how the transmission view handles the well-known cliff-top and bath-tub examples. On the transmission view of it is (roughly) the water flooding his lungs, or the fall off the cliff, that causes my uncle's death. My omission to save him is not me *causing* his death, it is me getting out of the way of what causes his death. (I can of course be morally responsible for this failure to be causally responsible.) The transmission view makes causation what it intuitively should be: a *local* affair. (As Dowe writes: "The process theory [of causation] is therefore *localist*, meaning that facts about causation hold in virtue of actual local features of the world" (2004, 927). This localist picture of causation as transmission of power goes closely with the local view of agency; getting one of these things right helps us to get the other right too.

Contrast the counterfactual theory's very counterintuitive distribution of causal responsibility, in principle, right across the whole universe at every moment. One classic argument for this sort of universally-distributed causation which makes the connection with the counterfactual account clear and explicit is Mill's in *A System of Logic* (3.5.3):

> If a person eats of a certain dish, and dies in consequence, that is, would not have died if he had not eaten of it, people would be apt to say that eating of that dish was the cause of his death. There needs not, however, be any invariable connection between eating of the dish and death; but there certainly is, among the circumstances which took place, some combination or other on which death is invariably consequent… the whole of which circumstances perhaps constituted in this particular case the *conditions* of the phenomenon, or, in other words, the set of antecedents which determined it, and but for which it would not have happened. The real cause is the whole of these antecedents… (Mill, 1872, 3.5.3).

Mill here attempts to combine a counterfactual view of causation with a regularity view. It is the regularity view that he wants to run as a *scientific* view about causation, one which is capable of giving us exceptionless instances of the form "As cause Bs." The counterfactual view is in tension with the regularity view, because it implies, as Mill does not explicitly say but seems uneasily to sense, that there can be no exceptionless true claims of this form at any less than a global scale. But a science that cannot generate usefully local and specific instances of "As cause Bs" is worthless. Science, in short, needs locality just as much as the theory of agency does.

The transmission model of causation implies that omissions cannot be causes. There is another obvious reason why omissions cannot be causes which is not proprietary to the transmission model of causation. This is that omissions are not just failures to prevent things that happen. They are also, by the same token, failures to prevent things that do *not* happen. My omission to check my car's tyres on January 1 is not just my failing to prevent January 14's actual accident. It is also my failing to prevent January 7's merely possible accident. My omission to send any donation to famine relief does not only mean that I fail to prevent a death in an African famine. It also means that I fail to prevent a death in an Ohioan famine—even though there is an African famine, and no Ohioan one. If causing, for omissions, is failing to prevent, then my omission causes not only the actual death in Africa, but also the hypothetical death in Ohio. This is an absurd conclusion.

Can we block this absurd conclusion by stipulating that nothing non-existent has a cause, so equating causing by omission with failing to prevent *something actual*? No. The whole point of the present objection is that omissions stand in exactly the same relation to the *actual* consequences that they fail to prevent as to the *merely possible* consequences that they fail to prevent. But patently, omissions do not stand in a causal relation to these merely possible consequences. Therefore, the omissions do not stand in a causal relation to the actual consequences either; those consequences are not *their* consequences. To stipulate that we are only interested in failures to prevent what actually happens is not to answer this objection. It is to decree that we shall ignore it.

In contemporary philosophy, the main culprit behind the mistaken view that omissions are causes is probably the popularity of the counterfactual model of causation. For this view of causation, and the denial of PDE (and AOD) as an immediate consequence of it, see Harris (1985, 29) (italics in original):

> *where something happens, or a state of affairs obtains because someone did something, I will say that the agent is positively responsible for its occurrence;* and *where a particular*

state of affairs obtains or something happens because an agent did not do something I will say that the agent is negatively responsible for its occurrence.

Here Harris conflates causal and moral responsibility, a conflation that Sect. 7.3 will show to be mistaken. But that is not the only thing wrong with his definition. Suppose there is a poisonous sea-anemone in a rock-pool on the coast of Thailand. Am I responsible for the anemone's existence "because" I have not flown out there to uproot it? Are you? Is everyone? It all depends on what we mean by "because"—a strikingly shifty word in Harris's argument. (In the absence of a theory of what it is to act, Harris's distinction between what an agent "does" and "does not do" is also highly ambiguous.)

It emerges in context that Harris has something very like the counterfactual analysis of causation in mind. But this model faces clear and decisive counterexamples.[13] One class of counterexamples have to do with overdetermination. Suppose the causal claim "John died because Crazy Joe the contract killer shot him" is true. But suppose too that there is another contract killer, Harry the Dog, lurking in the alley, who is just as competent (and crazy) as Crazy Joe, and who is also after John. In this case the counterfactual "If Crazy Joe had not shot him, John would not have died" will be false. Harry the Dog was also after the unfortunate John; so John would still have died even if Crazy Joe had not shot him. Whatever makes it true that Crazy Joe caused John's death, then, it cannot be the counterfactual that here is false.[14]

Again, even if there was a true counterfactual corresponding to every causal relation, there would still not be a causal relation corresponding to every true counterfactual. One sort of counterexample arises from identity relations: "If Napoleon had not fought at Waterloo then Bonaparte would not have fought at Waterloo." A second sort concerns relations in time: "If it was this time yesterday, I would be standing on the top of Ben Nevis." A third kind of counterexample is mathematical: "If π were exactly 3, mathematics would be a much easier subject." All these may perhaps be true counterfactuals, but none of them commits us to positing a causal relation between the states of affairs mentioned in the antecedent and in the consequent.

So for the counterfactual account of causation to be plausible, it must not be committed to the claim that counterfactuals are the very same thing as causal statements. But then what is it that distinguishes the counterfactuals that *are* causal statements from the ones that are not? This is a hard question for the counterfactual theorist. He cannot answer it by saying merely that they are the causal ones, since that is question-begging. He cannot answer it by giving a further account of what causation is, either: not, at any rate, so long as his position is that the counterfactual approach can give us a full and adequate account of causation all on its own.

Even if he can circumvent these problems, there are others. One is that the counterfactual model can't tell a cause from a background condition in any non-observer-relative way. Suppose these are all true counterfactuals: (1) "If John had not struck the match against the sandpaper the match would not have lit;" (2) "If there had been no oxygen in the room the match would not have lit;" (3) "If a meteorite had hit the town the match would not have lit." An adherent of the counterfactual theory is obliged to take (1), (2) and (3) as causal truths of exactly the same status objectively speaking. There being oxygen in the room, John's striking the match, and the non-occurrence of a meteorite impact must count for him as equally causes of its combustion. He is obliged to take the highly paradoxical view that it is nothing in the nature of the case, but only our subjective interest in one factor rather than another, that makes us call John's striking of the match the cause of its burning rather than the meteor's non-impact and the presence of oxygen.

Of course, many distinguished theorists of causation, from Mill and Honoré and Hart to Lewis, have swallowed this consequence. This is perhaps because they take it that the questions "What is the cause of X?" and "Why did X happen?" are the same questions. But, on at least some reasonable understandings of the latter question, this is a mistake. It conflates the interest-relative notion of *causal explanation* with the non-interest-relative notion of *giving the cause*. An example may help to bring out this distinction. If an experienced car mechanic asks "Why did the car crash head first into the wall?" he is not asking whoever answers to give the cause—namely, the momentum of the car. That is something he knows about already. Rather, he is asking for an explanation of what he doesn't yet understand: namely, what was different in the causal set-up this time around, as compared with some run of other similar occasions that interests him. This example shows that causal explanation is an interest- and information-state-relative notion: what counts as *causal explanation* to a given person depends on what that person already knows, and also on what the person is interested in. (The car mechanic does not wish to hear about the quantum physics of a car crash, either.) *Giving the cause* is not interest- or information-relative in this way: to give the cause of an event is simply to say what brought it about, without in any way tailoring this statement to what one knows one's audience already knows. We should be careful, then, not to conflate *causal explanation* with *giving the cause*. Especially when discussing action and omission; reflection on an approach like Harris's (or Rachels' in Rachels, 1975: more about him shortly) quickly leads one to the diagnosis that part of the problem with their discussions is that they do not keep this distinction clear.

The transmission model of causation shows us how easily we can avoid this conflation. Whatever we might wish to say under causal explanation (which, to repeat, will depend on what our audience already knows), consider what, on that model, we can say to give the cause of some particular event. John's striking of the match, i.e. his accelerating the match-head along the sandpaper strip on the side of the matchbox, transfers energy in the form of friction into the phosphorus head of the match, which (to abbreviate the full scientific account of combustion) responds by bursting into flame. In this story, does the oxygen in the room initiate a transfer of energy? No, though it does facilitate it: the oxygen is a medium for the burning of the phosphorus. And does the non-impact of any meteor initiate a transfer of energy? Of course not: non-impacts are non-events, and non-events have no causal powers at all. Thus the transmission model not only gives us a non-subjective way of identifying John's striking of the match as the reason why it burns. It also gives us a non-subjective way of sorting the *cause* of the burning, from a *medium* for the burning, from a *negative necessary condition* for the burning. These further distinctions are useful too.

A satisfactory account that gives the cause of the match's burning will make John's striking of the match obviously and objectively salient. It will not make salient the presence of the oxygen in the room or the absence of the meteor, even though the counterfactual is certainly true that, if either of these conditions were not as it is, the match would not strike. If these conditions were otherwise, then they would *become* objectively salient in the causal story; they might become important in causal explanation, even though they are not central to giving the cause. But that is no reason for saying that these factors are *already* salient before they are varied, just because varying them is possible. After all, there are indefinitely many such factors, and they can't all be salient. We have here another instance of the same asymmetry as before. In producing variants upon a basic model of causal story, we will find that the factors that are salient in the original version (e.g. John's striking of the match) remain salient in the variants, unless those variants introduce a special reason why the factors cease to be salient; while the factors that were not salient in the original version (e.g. the presence of oxygen, or the absence of a meteor impact) remain non-salient in the variants, unless those variants introduce a special reason why these factors should become salient.

Thus the counterfactual account of causation is hindered by its lack of clarity about the difference between causes, mediums, and necessary conditions; also by its conflation of causal explanation with giving the cause. And without a cumbersome appeal to a possible-worlds semantics (which notoriously brings troubles of its own), it cannot represent the intuitively important notion of differences in causal weight. If anything is a *conditio sine qua non*, it counts as a cause for the counterfactual account.

The counterfactual account also faces a problem that we might expect to crop up for a theory that (as I pointed out above) distributes causal responsibility right across the whole universe at every moment: it gives us too many causes. (Here too the connections between "global" views of causation and "global" views of action and responsibility are not hard to see.) If John's great-great-grandparents had never met, then John would not be around to strike matches now. So the counterfactual model of causation implies that their meeting is a cause of the match's igniting. More generally, the counterfactual model of causation implies that most of the history of the whole world up to now is the cause of every event now. But this is manifestly absurd, and the transmission account helps to explain why: to know the cause of an event you need to know something about the more proximate energy-transfers that brought it about—but not *everything* about *every* energy-transfer in its causal antecedents.[15]

I have argued in favour of the transmission model and against the counterfactual model of causation for two reasons. First, because I think the transmission model is right and the counterfactual model wrong. Secondly, because (as I said above) on the transmission model it is obvious that omissions are not causes, since causes are by definition transmissions of energy, and omissions never are.

But suppose you are undeterred by these difficulties, and insist on sticking with a counterfactual model of causation. Even then, it turns out either that omissions are not causes, or else that causes are not additive (that is: "A causes E1" and "A causes E2" do not imply "A causes E1 and E2"). To see this, compare two pairs of counterfactuals:

(A1) If you had not sent poisoned food to Africa Aid, ten deaths would not have occurred;
(A2) If you had not sent poisoned food to Afghan Aid, ten deaths would not have occurred;
(B1) If you had given £100 to Africa Aid, ten deaths would not have occurred;
(B2) If you had given £100 to Afghan Aid, ten deaths would not have occurred.

If you send poisoned food to ten people in Africa and send poisoned food to ten people in Afghanistan, it follows that you cause twenty deaths. But if you omit to send £100 famine relief to Africa Aid and omit to send £100 famine relief to Afghan Aid, it does *not* follow that you omit to prevent twenty deaths. For if you omitted to send £100 to each of these charities, you also omitted to send each of them £50; and you also omitted to send each of the £200, £400, £600. In this sense, there is *no* particular number of deaths that you fail to prevent: possibilia are not determinate enough to be countable in that way. So even the counterfactual theorist should not count omissions as causes. Omissions are not causes at all; hence my claim that omissions always score zero—at least on the causation test for actionhood.

But what about the other two tests? Can't there be an intentional omission, performed in a situation where the agent could have done otherwise than omit? And won't such an omission have some non-zero degree of actionhood? Indeed there can. To close this section, let me add a brief word on intentional omissions, and on one other matter arising at this point: whether there are any cases that are not neatly classifiable under the action-omission distinction.

Intentional omissions. As an example of an intentional omission—leaving some action undone precisely in order that a given state of affairs will ensue[16]—we may as well consider again the well-known bathtub case. Here it's often asserted (e.g.) that callous Chris, who allows his rich nephew to drown in the bath, is no better than interventionist Ian, who actively drowns the same rich nephew (Rachels, 1975 is the originator of this example). Such cases are supposed to point to the moral that actions do not differ in point of culpability from the corresponding[17] omissions. Since the action-omission distinction is only supposed to give us a *ceteris paribus* moral truth, such a general moral should not be inferred from a single example, or even from a range of examples, unless there is reason to think that these examples capture a general pattern. Of course omissions and actions are alike in culpability sometimes, and in particular, when both are intentionally performed by agents who desire the same reprehensible outcome.[18] But that is no objection to the present thesis that sometimes they are *not* alike in culpability—as when an agent chooses to omit to prevent a death that he does not wish to see happen, but thinks will happen, where he could also have chosen to act to cause it.

Unclassifiable cases. Nor is it an objection to the act-omission distinction to cite the fact, if it is a fact, that not every case of behaviour can be easily sorted into the category of act or omission. The distinction does not claim this, but only that it is *usually* clear whether something is an act or an omission, and that where it is, its status as act or omission is a feature of it that can be (not: *must* be) relevant to assessing actionhood and responsibility.

And *is* it a fact? The usual examples of allegedly borderline or unclassifiable cases include the safety net case discussed by McMahan and Howard-Snyder (McMahan, 2002; Howard-Snyder, 2002): removing the safety net seems to be killing the person who would have fallen onto it had it been there, and now falls onto some lethally hard surface instead. But the agent who removes the safety net does not *act on* the faller, in the sense of transferring energy directly to the faller's body. So it looks like removing the safety net is allowing the faller to die, not killing her. What is wrong with that conclusion? The people who object to it usually, I think, are worried by the thought that when someone is murdered, there must be a single causal process involving the murderer and the victim. But that thought is mistaken anyway on the transmission theory of causation, which straightforwardly implies that it is possible for A to murder B without energy being transferred directly from A to B, and so without A causing B's death. We should not make too much of a meal of this implication.[19] It will only sound paradoxical as long as we do not notice the difference between moral responsibility and causal responsibility. The failure to keep these two concepts apart is the bane of too many discussions of this subject. There is nothing very surprising about the thought that, in general, murders can be committed by negligence or other forms of omission[20]: not because the murderer's deliberate negligence imputes *causal* responsibility to him, but because it imputes *moral* responsibility to him, on the basis of his deliberate or negligent failure to be causally involved (in the crucial way, at the crucial time). This model captures what is going on in the safety-net case. Notice, too, that there are plenty of ways of spelling out the safety-net story so that it is not a murder; perhaps the agent who moves the safety net is too young to be responsible, or is trying to prevent three other fallers from dying, or simply gets his calculations wrong. These factors vary moral responsibility without making any difference to causal responsibility.

So safety-net cases are not unclassifiable by the action-omission distinction; they are straightforwardly cases of failure to prevent, where this failure may or may not be culpable. Are there other cases which look harder to classify as actions or as omissions? It would be unsurprising if there were, and entirely unthreatening to the action-omission distinction; nothing in the nature of that distinction entails that there must always be a hard line to be drawn between omission and action, or that there cannot be indirect ways or marginal or borderline cases of transferring energy, and so of acting. Having said that, many supposed examples of borderline cases are not really any more borderline than safety net cases. For instance the class of examples where a death happens because an obstacle to a lethal force is removed are not borderline in this way. Releasing a rock to fall on someone's head, for instance, is not a marginal case: it is straightforwardly a case of action. To call it an omission to prevent death, rather than a causing of death, would be like saying that a shooting is an omission to prevent death. After all, when I shoot someone, what happens is that I remove an obstacle to the release of the energy in the explosive charge.[21] It hardly follows from this that shooting someone is an omission rather than an action. Shooting someone is as plain a case as there could be of the transfer (the direction or redirection) of energy, and so of action. Releasing a stonefall is not much less plainly a case of the same thing. When I shoot someone, I aim the bullet at him; when I release a stonefall intending it to kill someone, I aim the stonefall at him. When I remove a net from under someone who falls, I do not, in the same way, aim *him* at the ground, because I am not part of what causes him to fall.

Cases involving life-support machines do not look much more genuinely borderline, either. Suppose we have one machine that keeps going unless you press an Off button, and another that shuts down unless you press a Continue button every 24 h. We might think that there is no real difference between action and omission here: "You're just pressing a button." But this is a confused response to the case. It plainly *is* possible to distinguish the action of pressing the button from the omission to press it: so the action-omission distinction does apply in this case.

Of course you might also say—that this is a different point is what makes the "You're just pressing a button" response confused—that although the action-omission distinction can be drawn here, it is hard to see that it has any moral significance in this case: "After all, the technical details of how a machine is designed cannot matter much morally speaking." Certainly there are cases where the action-omission distinction can be drawn, but makes no serious moral difference: perhaps Rachels's pair of bathtub murders, for instance (Rachels, 1975). But it is not obvious that the life-support machine is such a case, and there are hard-thinking professional casuists whose view is that it makes a huge moral difference which way the machine is designed. (The Orthodox Jewish view is that life-support machines must have a Continue button, not an Off button, precisely so that if deaths happen, they will happen by omission, not action.)[22] Here too, then, we have a case which does not seem to be borderline between action and omission.

To sum up this section: actions and omissions are both things for which we can be morally responsible; but we cannot be causally responsible for the consequences of omissions, because omissions have no causal consequences: omissions are not causes at all, but failures (deliberate or not) to intervene causally. On the causation test of actionhood, therefore, omissions score lower than actions; omissions likewise tend to display a lower degree of responsibility or of culpability than actions do (though there are exceptions: there can be fully deliberate omissions). And this establishes the principle about action and omission that I am arguing for: that *ceteris paribus*, we are less responsible or culpable (and possibly not responsible or culpable at all) for our omissions than for our actions. In Sect. 7.4 I turn to the parallel principle about intending and foreseeing.

7.4 Spelling Out the Simple Strategy: (7), (8), (9), and PDE

Step 7 of the simple strategy of argument, as I presented it in Sect. 7.1, said this: "*Ceteris paribus*, what is foreseen but not part of an agent's plan either as end or as means is less intended than what is part of the agent's plan either as end or as means, being only (i) foreseen or (ii) unconsidered." Step 8 inferred, from Step 7 and Step 3's claim that intendedness is one of the tests for actionhood, that actions have a lower degree of actionhood under (i) unintended but foreseen descriptions than they do under intended descriptions, and a lower degree still under (ii) unintended and unforeseen descriptions. Then Step 9 drew the conclusion that vindicates PDE—that degree of intendedness can make a moral difference.[23]

Turning to the detail of this part of the argument: Step 7 is hard to defend only in the sense that it is so obviously true that it is not entirely easy to see what argument it needs. If I mean e.g. to buy a train ticket so as to travel to London, then obviously I foresee that Joe, the stranger in front of me in the queue at the ticket desk who (as I overhear when he buys his ticket) is going to London by the same train, will be travelling with me; without it being in any sense my *plan* to travel with Joe. Of course, I might have struck up a conversation with Joe in the queue and decided already that I like him (or alternatively that I have a good chance of defrauding him of his life savings during the journey); in which case I may well form a new plan that *will* include travelling with Joe. But these are different possibilities, involving different plans, from the case where I know that Joe goes too, but have no specific intentions one way or the other about him.

It can't seriously be disputed, so far as I can see, that there is a difference between the category of what I plan, and the category of what I foresee but do not build into my plan. Can it be argued, though, that while the latter category is real, there is nothing in it? The idea would be that, though we can in principle make this distinction, in practice it doesn't apply to our planning. *Everything that I know will be changed by my action* is part of my plan, so that the category of what I foresee but don't plan is empty.

Perhaps this claim will be advanced, not as a commonsensical one, but as a purported philosophical improvement on common sense: "This is how we *should* reason." Perhaps that's how Henry Sidgwick meant it. Sidgwick writes:

> *it is best* to include under the term 'intention' all the consequences of an act that are foreseen as certain or probable; since it will be admitted that we cannot evade responsibility for any foreseen consequence of our acts by the plea that we felt no desire for them, either for their own sake or as means to ulterior ends: such undesired accompaniments of the desired results of our volitions are clearly chosen or willed by us (Sidgwick, 1874, 202, italics added).

Here there seems to be an ambiguity in Sidgwick's use of "chosen." Suppose in desperate circumstances I attempt to rescue a skater who has fallen through the ice, knowing that very likely I will fall through too, but seeing no better alternative to my gravely perilous rescue attempt. In this case, Sidgwick will say that *my fall through the ice is something that I have chosen*, meaning that I freely went into a situation where I knew it was very likely to happen. This is not in fact something that I think we should say; but it is not implausible for Sidgwick to suggest that we might say it. However, it seems very implausible indeed to add as an inference from that—as Sidgwick here commits himself to doing—that *I have chosen to fall through the ice*, meaning that it *was my intention to* fall through the ice. For it isn't even plausible to say that I fall through the ice *intentionally.*

There is a simpler and more fundamental reason why Sidgwick's main proposal in this passage is mistaken. This is because of the falsity of the global view of action. We do very few things indeed with an overall resultant state of the world in mind. *Some* actions might be like this, such as pressing the nuclear button (or pressing another button that overrides it). But actions of that sort are pretty rare, and the contrast with more normal actions is obvious and revealing. For typical actions, it is nonsense to suggest that they are done with an overall state of the world in mind. Here I don't mean by "nonsense" just "a view that is very unlikely to be true:" I mean that I can't see how to make sense of the suggestion. Whenever I play any move at chess, must I keep an eye on the effects (and indeed the *possible* effects) of that move on politics in Kazakhstan? When I help my daughter with her violin practice, must I think about how this deed will impact on contemporary ethical theory? The answer to these rhetorical questions is "Patently, no." But the global view of agency tells me that in each choice I should be considering not just *anything*, but *everything*—Kazakhstani politics *and* contemporary ethical theory *and*... everything else there is to think about. This is simply insane. Contrast the "localist" view that, in making a move at chess, my plan is (say) to fork my opponent's rooks in three moves' time by making a pawn sacrifice now, while I foresee, and am prepared to tolerate, but do not intend, the result that my opponent will lose his temper when he loses one of his rooks, and neither actively foresee (i.e. actually think about) nor intend indefinitely many other results. In contrast, this view sounds like the heart and soul of sanity. But this is (what we can call) the local view of agency, and it is tailor-made to go with Step 7.

Not only are the global view of agency and monism about the good made for each other, as noted above; it is also true that the global view of agency seems to fit neatly with a counterfactual view of causation. That is, the global view goes nicely with the idea that we are equally responsible for *everything* that would be otherwise if we did x. Contrast the view that I have developed, that causation is transmission of energy. I have shown how the transmission model of causation helps us to vindicate AOD; the transmission model also has something to say in favour of PDE.

This becomes clear when PDE faces up to a familiar question: "How can you tell a means from a side-effect?" Suppose, e.g., that my plan of action is to knock you over to prevent a chimney falling on you, and that I foresee but do not intend that I will break your wrist in the process. Then compare:

(i) I intentionally prevent the chimney falling on you by knocking you over

with

(ii) I intentionally prevent the chimney falling on you by breaking your wrist.

PDE implies that (i) is true and (ii) false. What if anything justifies this verdict? The answer could hardly be simpler. The verdict is right because (i) is my *plan of action*, and (ii) isn't.

"But knocking you over *is* breaking your wrist; so plan of action (i) *is* plan of action (ii)." Not so, for two reasons. First, because what I need to do is to transfer the energy into your body that will get it out of the way of the falling chimney; this is a truth about the causal route that I need to go down. Do I need to transfer the energy into your body that will break your wrist? Extensionally yes—because as things are, my knocking you over is my breaking your wrist—but intensionally no—because what it is that I need to do in other close possible worlds is not *to break your wrist* but *to get you out of the way*. The causal route I go down in knocking you over tracks getting you out of the way in these other worlds, not breaking your wrist.

Secondly, if the identity of knocking you over with breaking your wrist implied that the two plans of action (i) and (ii) were the same, then we would by the same token have to agree that plan of action (i) is identical with *any and every plan of action at all* that, in its "by..." clause, mentions something (known to me to be)[24] extensionally equivalent to knocking you over. But this is obviously false: the notion of a plan of action is an intensional notion. This is part—but only part—of the reason why, as already argued, plans of action are local, not global.

"But if knocking you over is *bound* to produce the breaking of your wrist, then causing it to be that the chimney does not fall on you cannot reasonably be distinguished from causing it to be that your wrist is broken." But I've just shown that this is straightforwardly false: it can be distinguished intensionally, and there's nothing unreasonable about so doing.

Perhaps the counterfactual model of causation is at work here too, encouraging the sort of thinking behind this last protest. But in truth this consequence doesn't *obviously* follow even on the counterfactual model. These two counterfactuals are different:

(iii) If I had not knocked you over, then the chimney would have fallen on you.
(iv) If I had not broken your wrist, then the chimney would have fallen on you.

On the counterfactual model, (iii) and (iv) are both true, certainly. How is that supposed to show (if this is what opponents of the means/ side-effect distinction are trying to show) that knocking you over = breaking your wrist (or that intending knocking you over = intending breaking your wrist)? Remember that in general indefinitely many counterfactuals are going to be true for each event; there's no reason to suppose that the events mentioned in the antecedents of all of these are all identical.

But perhaps there is a thought in the offing that *does* force the counterfactual model to treat breaking your wrist as a means, namely this one:

(v) If I had not knocked you over, then I would not have broken your wrist; and if I had not broken your wrist, then the chimney would have fallen on you.

(v), we might say, describes a causal T-junction: knocking you over causes it to be that the chimney misses you down one branch, and that your wrist gets broken down the other. Yet for the counterfactual model of causation (v) is structurally indistinguishable from a linear three-term causal chain such as (vi) describes:

(vi) If I had not lit the blue touch-paper then the rocket would not have taken off, and if the rocket had not taken off, then it would not have broken the Ambassador's window.

Maybe it is *this* fact that encourages proponents of the counterfactual model into the error of seeing side-effects as no different from (no less causally relevant to a given end than) means. But the correct verdict on the counterfactual model's inability to tell the structure of (vi) from the structure of (v) is, surely, "So much the worse for the counterfactual model;" for once again an available *pari passu* move admits an absurdity. If the side-effect of my breaking your wrist is admitted into the causal story for the sort of reason that the comparison with (vi) suggests, then so, for the same reason, must any and every side-effect be admitted. And this quickly leads to the admission of indefinitely many intuitively quite irrelevant counterfactuals as parts of the full causal story. Which is absurd—and more evidence against the counterfactual model.

In defining what I directly intend, as opposed to what I foresee but do not intend, as whatever comes into my plan of action as a means or an end, I make the notion of direct intention an intensional one. It follows from this that I can simply skip familiar problems about "causal closeness." On an intensional story about plans of action, not even an identity relation between the causal factors A and B is enough to secure that both are parts of my plan in just the same way, since my plan can be about A (*de dicto*) without being about B. *A fortiori*, then, A's overlapping with B, or being very closely connected causally with B, can in no way secure this—*however* close the relation.

One consequence of this is that the name "principle of double effect" turns out to be a misnomer, as does "side-effect." Sometimes the "side-effect" that concerns us won't be a side-*effect* at all, but the circumstance that the very same event can be variously described, where we offer different moral responses to these different descriptions. What the agent must do about these alternative descriptions is captured by the proportionality clause.

This gives us an answer to the following question: "Isn't there something absurd about saying that you intend to silence someone by putting a bullet through his brain, but not by killing him?" Perhaps there is, yes. But if so, it has nothing essential to do with PDE. PDE as I'm defending it makes it perfectly possible that such a statement might accurately capture someone's plan of action. What would be morally wrong with someone who sincerely said such a thing could in principle easily be, *not* that he was making an arbitrary or implausible division between a means and a side-effect, but rather his failure to take proper moral account of the person's death *as* a side-effect. PDE as applied to killing gives us *pro tanto* moral permission, *not* to do absolutely anything that we can describe as a means-end action-plan involving no killing, but rather to do what we can describe as a means-end action-plan involving no killing *and with no unacceptable side-effects.* The moral fault involved in saying that you intend to silence someone by putting a bullet through his brain, but not by killing him, is not necessarily that killing him can't be called a side-effect—given the intensional individuation of plans of action, it can—but that this side-effect ought to appear to the speaker as obviously disproportionate to the good to be achieved by silencing the person. It's not as if we can disingenuously redescribe any doing in any way we like; we are morally responsible both for the way we describe things as means or side-effects, and for taking or failing to take seriously the moral significance of side-effects.

"Then haven't we replaced the old problem about causal closeness with a new problem about descriptive closeness?" Yes. But it's not *such* a problem. As I just said, it's part of the agent's moral responsibilities to spot such relations of descriptive closeness.

"But this is a recipe for all sorts of special pleading!" If this is the charge that I am telling agents disingenuously to (re-)describe their deeds in ways that make them appear morally licit when they are really known not to be, then I deny the charge. I am not advocating moral insincerity. The fault of moral insincerity is one that I take as seriously as anyone, but that fault has nothing specially to do with the issue whether PDE is true or false. Or if the charge is that I am encouraging agents to be *obtuse* about what other descriptions of their deeds are available and applicable besides those under which they actually do them, then again I deny the charge: PDE has a proportionality clause in it, and that clause does important work.

"No; it does *all* the work. For the way you talk, just any action *could* pass the test of involving nothing illicit in the specification of its means and its end. Everything then hangs on whether the action passes the proportionality test. And so you are heading towards consequentialism." This objection involves a *non sequitur*, since the proportionality test is not a consequentialist test. Consequentialism means assessing rightness and wrongness on the basis of consequences alone, but PDE's proportionality clause is only part of how PDE tells us to assess rightness and wrongness, and is not restricted to consequences. As pointed out above, some of the "side-effects" that will need to be considered under the proportionality clause are not *effects*, but alternative descriptions, of a proposed action. The objection also seems to involve a factual mistake: it seems to suppose that agents do not in fact ever set out with the express intention of e.g. murdering or torturing someone. But real agents form such intentions all too often, as indeed do agents in philosophical examples.

It's vital to a correct understanding of PDE to see that the agent's own account of his intentions is normally authoritative. This point crucially goes missing in many discussions of the PDE, e.g. Warren Quinn's redescription of the PDE as a distinction between direct and indirect agency. Of course self-deception about what my own motives and intentions are is possible. That aside, *the agent is authoritative about what he intends; and what he actually intends is the key.* "What he *actually* intends:" to repeat, I am not saying that if action-type A *can* be done with permissible intentions, then any A-type action is fine.

Measure against this point what e.g. Alison McIntyre writes in the *Stanford Encyclopedia of Philosophy* entry on "Double Effect:"

> . . .if the soldier who throws himself on the grenade in order to shield his fellow soldiers from the force of an explosion acts permissibly, and if the permissibility of his action is explained by double effect, then he must not intend to sacrifice his own life in order to save the others, he must merely foresee that his life will end as a side-effect of his action. But many have argued that this is an implausible description of the soldier's action. (McIntyre, 2004, section 3).[25]

Maybe they have; but are they right? It is perfectly conceivable that an agent might report his own intention as this: to "throw himself on the grenade in order to shield his fellow soldiers."[26] Where an agent thus reports his own intention, and where there is no reason to think that he is self-deceived or dishonest, *others are in no position to doubt his report.* In any case, with this story we are considering an imaginary scenario in which it was *stipulated* that the soldier's intention was to "throw himself on the grenade in order to shield his fellow soldiers;" so again we can ask what exactly McIntyre supposes is bound to be wrong about that stipulation, or indeed how anyone, in the circumstances she describes, could possibly be in a position to find it "implausible."

A little later McIntyre says this:

> Those who say that it would be impermissible to perform an abortion to save the life of a pregnant woman say that this is because this would involve intending the death of the fetus. However, if it is also maintained that a hysterectomy may be performed on a pregnant woman with uterine cancer because the death of the fetus would be a merely foreseen side-effect of surgery, it is hard to find a principled ground for drawing this distinction that could serve as a guide to moral judgment (McIntyre, 2004, section 3).

On the contrary, it isn't hard at all. The principled ground is, once more, simply the agent's intention. As a matter of plain fact, doctors performing abortions typically do intend the death of the foetus, as they will tell you (reluctantly perhaps) if you ask them. Nor is there is anything impossible about a doctor's having an intentional plan "to cure this woman's cancer by performing a hysterectomy on her," without the death of the foetus in the uterus being more than a side-effect that the doctor foresees but does not intend. As before, the agent is (normally) authoritative about what his own intentions are. Also as before, the question what intentions an agent may permissibly have needs to be kept sharply separate from the question what side-effects of what actions an agent may permissibly tolerate; which is why, though it would of course *also* be possible for a doctor to perform an abortion sincerely not intending but only foreseeing the death of the foetus, that fact does not force an opponent of abortion to concede that abortion is licit if hysterectomy on a pregnant woman is licit.

In this section I have clarified the distinction between means and side-effects by insisting on the intensionality of intentions; and I have clarified the answer to the question "What in fact does an agent intend?" by insisting on the first-person authority of the agent, except in very special circumstances, to report his own intentions correctly. These clarifications get us past the main objections to PDE. Since PDE has (as I've also argued in Sect. 7.4) a good deal of intuitive support, this means that the truth of PDE is reasonably safely established, alongside the truth of AOD, as argued for in Sect. 7.3.

In Sect. 7.5, I round off my discussion by looking at two famous problem cases relevant mainly to PDE, but also, to a lesser extent, to AOD.

7.5 Trolley, Transplant, and Demandingness

First, here is the well-known Trolley case:

> *Trolley.* 5 miners will die if a trolley goes down to the end of Track A; 1 miner will die if it goes down to the end of Track B. So Mike, who is trapped on the trolley and can do nothing except determine which track it goes down, sends it down track B, saving 5 miners at the cost of 1 life.

Trolley on its own is not a serious problem for PDE as I defend it. My intensional account of how to individuate plans of action entails that Mike's plan of action can[27] be "To save five miners by diverting the trolley onto a track with one miner on it." This plan of action involves no intention to kill, and even if it involves a foreseeable death, this bad effect is plausibly proportionate to the good effects of Mike's action. Indeed, Mike would not necessarily intend to kill even if he diverted the trolley towards the five miners and away from the one: his intention in that case could still be "To save one life by diverting the trolley." However, this plan of action would be criticisable under proportionality; it would compare very unfavourably, as to the side-effects that it commits Mike to tolerating, with the option of doing it the other way round and saving five lives. (Compare the familiar case of the crashing aeroplane, in which it's obvious that the pilot of the stricken airliner should steer so as to crash in the countryside, not on the town).[28]

Contrast the Loop case, where a different trolley in a different mine can only be stopped from hitting five miners (going down Track C) by running it into the body of 1 miner who has somehow become trapped across the rails in such a way as to create an obstacle on Track D, which loops back onto Track C. What we should say about Loop depends again on the agent's precise intentions. If his intention is to save the others *by killing this miner*, then PDE says that what he does is straightforwardly wrong, because he is killing one miner as a means to save the others. But if his intention is to save the others *by running the trolley into the miner's body*, with the miner's death as a foreseen but unintended consequence, then PDE does not rule out his action. If there is something wrong with what he does, it is not that he illicitly kills an innocent as a means to some good. For it is not his intention to kill an innocent—though it may be his intention to assault an innocent's body as a means, which is probably also illicit. Nor is the action criticisable because it breaches the requirement of proportionality either; after all, more miners are saved this way.

What about the transplant case that is so often discussed together with Trolley (the *locus classicus* is Thomson, 1985; cp. most recently Shaw, 2006)?

> *Transplant.* Five patients need life-saving transplants; one tramp passing by has all the organs they need. So Mark, the attendant surgeon, kidnaps and dissects the tramp, saving five patients at the cost of 1 life.

The comparison between Trolley and Transplant seems to set up a tricky technical challenge, since our intuition is that what Mark does is monstrously wrong, whereas what Mike does is permissible. Yet Transplant is very hard to distinguish, with respect to PDE, from Trolley (and even harder to tell apart from Loop). In Transplant, Mark can appeal to proportionality, just as Mike can in Trolley or in Loop. And it is possible for Mark's intention to be, not to kill, but to save the lives of the other patients by cutting into the tramp's body: just as Mike's intention in Trolley can be, not to kill (nor even to assault), but to save the lives of the other miners by diverting the trolley, and in Loop, not to kill (nor even to assault), but to save the lives of the other miners by using the body of the trapped miner to stop the trolley. This seems to suggest, very counterintuitively, that provided Mark does not intend the tramp's death as a means to saving the five patients, rather than merely foreseeing it, his action can be licit if what is done in Trolley or in Loop can be licit.

It is no answer to this to say that Mark typically *will* intend the tramp's death. That seems to be true, insofar as the word "typically" can be applied to such a fantastical case. However, it does not touch the problem as to what we should say is the moral distinction between Mark and Mike when Mark does *not* intend the transplant victim's death, but merely foresees it. Nor is the answer (*pace* Shaw, 2006) that Mark must intend to do something morally illicit, e.g. an assault, even if he does not intend to kill. It simply isn't true that Mark *must* intend that. As noted in Sect. 7.4, intentions are intensional entities, and as such, can be "sliced" as thin as intensions can—certainly thin enough for it to be possible to intend cutting without intending killing *or* assault. Moreover, Mark is normally authoritative about what his intentions are.

Nor is the problem that Mark will be mad if he intends organ removal without intending killing. Similarly shaped intentions in other cases are deemed by PDE not to be mad (or bad) at all: the whole point of applying PDE in bioethics—some would say—is that it enables us to explain e.g. how one can intend hysterectomy on a pregnant woman without intending the death of the foetus, or how one can intend pain relief for a terminally ill patient without intending the death that it is known will result. So in the case where Mark does not intend to kill or mutilate the tramp by dissecting him, but only to procure some organs for transplant, it seems that PDE does not explain what's wrong with what he does.

Nor will it do to say that we should apply AOD, not PDE, to Transplant, while applying PDE, not AOD, to Trolley. This seems simply arbitrary: what is the rationale for treating Transplant under AOD but Trolley (and Loop?) under PDE? Why not do it the other way round, and treat Trolley under AOD (giving the answer that it's better not to intervene positively, because this will be actively causing one death rather than letting five happen) but Transplant under PDE (saying, in effect, that you can foresee but not intend as many tramps' deaths as you like provided this leads to net life-savings)?

There might seem to be a general problem about arbitrariness here, of course. Why in general, someone might ask, do PDE and AOD get applied separately to solve separate problems? Perhaps a sense that it would lead to arbitrariness to keep both distinctions in play is the reason why so many philosophers try to get by with only one of them, or to reduce the one distinction to the other.

This question involves a misunderstanding. As was already implicit in **3**, the answer is that they *don't* get applied separately. They work together, as different tests that both apply to the same actions (or other behaviour). Thus there is nothing in AOD to stop us distinguishing, or to tell us that it is not morally significant to distinguish, intentional and unintentional omissions. Where we are talking about unintentional omissions, AOD tells us that *qua* omissions these generally need little moral defence, unless there's a special reason why we *should* have done something; and PDE tells us that the same verdict applies to them *qua* unintentional. With intentional omissions, it is even clearer that these are fitted for the simultaneous application of PDE and of AOD. So in Trolley, what's wrong with doing nothing to divert the trolley is that this intentional omission has foreseeable unnecessarily bad consequences, whereas what's *not* wrong with doing something is that (as PDE shows) this intentional action will not be a choice to kill. AOD does not enjoin omission rather than action, because the action in question does not have to be one of an impermissible type. PDE does enjoin acting in a way which will minimise deaths if this can be done without intending deaths—and PDE says it can. So this solution to Trolley is not an application only of PDE only, but of AOD as well; it is just that AOD does not rule out either diverting the trolley or not diverting it, whereas PDE does rule out not diverting it. The two distinctions are not being applied separately here, but in combination. This one example stands for a general pattern.

How, then, should we answer the main question I have been asking in this section—namely, "what is wrong with Mark the surgeon's action in the variant of Transplant where Mark does not intend the tramp's death, nor even an assault on him, but simply the removal of some organs from his body?"? The answer, I think, does not depend in a direct way upon either PDE or AOD, but (once more) upon the difference between the local and the global conceptions of action that both those distinctions presuppose. To put the point at its simplest, Mark's conception of action is not local enough: he is too ready to consider drastic interventions which drag into the frame of the moral problem others who are outside that frame, and have a right to remain outside it. Contrast the Trolley case, where the miners and Mike, the man on the trolley, are already within the frame of the moral problem whether they like it or not.

And what makes it true that a given factor, or person, lies inside or outside the frame of a given moral problem? The answer is hard to spell out in precise detail for particular cases. But that does not make the distinction between the "inside" and the "outside" of a moral framework any the less real. However we draw it for each case, there is a moral distinction between being and not being in an emergency, and hence between the moral demands that may legitimately be made on those who are and those who are not in emergencies. The difference between the lone miner's position in Trolley and the tramp's position in Tramp is one application of this distinction.

To add some intuitive backing to this distinction, we can point out how deeply it is ingrained in commonsense ethics. Consider the policy implications as we might call them—the broad logical consequences—of assenting to Mark's and Mike's choices. The wider implication of Mike's choice is that each of us will have to assent to the following general principle:

> (P1) If I find myself in a desperate situation where I and others face a lethal threat which, if it does not kill me, will kill more people also in the situation, then I cannot reasonably object to not having my life preserved.

By contrast, the wider implication of Mark's choice is that each of us will have to assent to the following general principle:

> (P2) If anyone finds himself in a desperate situation where he or she faces a lethal threat such that fewer lives will be lost if I am killed, then I cannot reasonably object to not having my life preserved.

(P1) is a reasonable principle because it restricts itself to putting my life in jeopardy in cases where I am already involved—emergencies *that I am actually in.* This is a genuine restriction on (P1)'s scope, because most of us are not in that many emergencies during our lives. (P2), by contrast, is a wholly unreasonable principle, because it includes no such restriction: (P2) says that my life is in jeopardy *anywhere* where killing me will bring it about that more lives are saved than lost. (P2)'s scope is almost universal, because there are indefinitely many possible cases where killing me *could* bring it about that more lives are saved than lost: the number of such cases is limited only by our imaginations, and our moral scruples. The practical upshot of (P1) is a society where a case like the Trolley would lead to Mike's being put on trial for murder, but then acquitted. The practical upshot of (P2) is a society where not just tramp-abductions, but all sorts of high-handed murderous interventions, happen absolutely all the time, and can happen to absolutely anyone—to the very reasonable fear and despondency of the populace.

(P2) also raises the fascinating practical question "*To whom* are we going to give the unlimited power to make these redistributions of life, liberties, organs, and so forth?": a question to which I see no answer that does not entail a sinister authoritarianism.

As a result, any reasonable consequentialist will accept (P1) and reject (P2) just as decisively as a non-consequentialist like me will. The difference between us will only be that the consequentialist will have to justify this choice simply because it is what fits with popular opinion; whereas the non-consequentialist can justify it by appealing to the local conception of action.

This line of argument thus brings it out one more time that local conceptions of action, causation, and responsibility are all made for each other—as are the corresponding global conceptions. The notion of the plurality of the goods is connected too. It is because the world is full of different kinds of values that we get this kind of locality, which would surely not occur if there was merely one kind, aggregable across all circumstances and times.

Notes

1. Apologies for the neologism.
2. It is important to distinguish moral from causal responsibility. But here, I work with an undivided notion of responsibility.
3. Arguably (cp. Note 4) there are mental actions—thoughts, judgings, directings of attention. Are these counter-examples to the supposed platitude? I see no reason to think that; only a hardline dualist need say that thoughts are not in the world. And I am not a hardline dualist. (I am not even a dualist, for a reason like the reason why I am not a naturalist: I do not know what the mental/ physical contrast is supposed to be, nor why it should be thought exclusive.)
4. Formulation retained from my statement of the Symmetry Thesis. The formulation does not, and is not supposed to, imply that actions *by definition* consist in or even supervene on bodily (non-)movements. Plainly they don't, even in the weakest and most "global" sense of supervenience. Since (attentive) thinking is action, and since thoughts do not necessarily influence overt behaviour and are not necessarily communicated, there can be a difference in the actions that it is right to ascribe to two agents without there being any difference in how their bodies move or do not move at any stage in their lives. Mental acts are not *usually* this private, of course—and naturally, in such a case no one could know whether their action-ascriptions were right or wrong. But they *can* be this private. Which is enough to show that action does not *in principle* supervene on bodily movement—even though a lot of action does *in fact* so supervene.
5. The ski-jumper's keeping still *may* be a negative action—though it may also be his omitting to do what he has a special responsibility to do, namely deflect the angle of his jump so that he misses the president. Even if it is a negative action, it does not seem to be the main thing that he is culpable for in the situation. His primary responsibility is for the movement that initiates his murderous ski-jump, which is not a negative but a positive action.
6. Here there arises a strong temptation to treat anything that involves moral effort as an action, as well as anything that involves physical effort. Provided we keep clear on the difference, there is no particular need to resist this temptation.
7. Philosophers can of course hold that omissions are causes, and still hold that there is an asymmetry between omissions and actions as to the causal relation that they stand in to their consequences: see, e.g., Sartorio (2005).
8. Isn't "transmits" already causal language? No: it means only that energy-quantity X was attached to object Y at time t1, and then is attached to object Z at time t2.
9. As Wesley Salmon more formally expresses it: "A causal interaction is an intersection of world-lines that involves exchange of a conserved quantity" (Salmon, 1997, 468).
10. This is not to say that there is nothing going for the regularity theory of causation—nor for the other three theories criticised in this paragraph. Different accounts of causation often seem, in fact, to be answering different questions.

What the regularity view does well, as emphasised by Pettit (1986), is capture the "regularising" aspect of causal explanation—the sense in which a fully articulated story of how something happened reduces that possibly surprising happening to a background of familiar regularities. (As Christopher Martin has emphasized to me in conversation, the doctor in Molière who explained a patient's drowsiness by a drug's *virtus dormitiva* was offering a perfectly good regularising explanation; the only thing wrong with it is its brevity.)

11. Raising the chance of the effect is certainly not sufficient for being a cause: there being oxygen in the atmosphere will raise the chances of all sorts of effects of which that circumstance is not the cause, but a precondition. Nor, apparently, is it necessary. Suppose I am a student looking for a wife, but I won't marry an Abba fan. Then the University registrar doubles the size of my previously Abba-fan-free class by admitting 20 new female students, 95% of whom are Abba fans. The chances of my finding a wife in my class have gone down in proportion to the 42.5% increase in Abba fans in my class. Yet the one girl in this batch of 20 who *isn't* an Abba fan turns out to be the one I marry. So the registrar's action *both* lowers the chances of my finding a wife in my class, *and* causes me to find a wife in my class. (You can of course object to this that the registrar's action only lowers the chances of my finding a wife in my class when we describe it a certain way, viz. as increasing the proportion of ineligible women in my class; described another way, e.g. as introducing a girl into the class whom I will find magnetically attractive, the registrar's action raises the chances. But this is not a problem for my counter-example to the probabilistic account of causation. It is a problem for the probabilistic account, the problem of the description-relativity of all probability-statements.)
12. It is for instance discussed in Howard-Snyder (2002, section 7), though the only source she cites for the "transfer of energy" view is conversations with John Hawthorne.
13. Parallel problems, and parallel counterexamples, face Mackie's account (Mackie, 1974) of causes as INUS conditions of their effects (insufficient but necessary parts of unnecessary but sufficient conditions).
14. Modify this case a little, and it may give us another counter-example to the raising-the-chances account of causation. Suppose Harry the Dog is an even more competent and even crazier contract killer than Crazy Joe. Then John will be even likelier to die if Crazy Joe does *not* try to kill him, but steps aside and lets Harry the Dog do it. So Crazy Joe's *not* shooting John does more to raise the chances of John's death than Crazy Joe's shooting John, hence is more of a cause of John's death. This conclusion is most likely absurd, and in any case awkward.
15. The counterfactual model might also seem to imply the absurdity that we cannot characterise the difference between an immediate cause and a distal cause. The argument would be this: In ordinary causal discourse we want to say things like "A caused B caused C caused D." On the counterfactual account, C's causing D means that there is a true counterfactual "If not C, then not D." But if A caused B which caused C which caused D, then on the counterfactual

account there will also be a true counterfactual "If not A, then not D." So we might just as well say that A caused D as say that C did. In other words, the counterfactual account of causation cannot represent the difference between a diachronic causal chain and a synchronic combination of causes. However, this objection overlooks the possibility that the counterfactual model might treat immediate causes as those counterfactuals of the A-D form which are true when there is no true counterfactual of the A-B, B-C, or B-D form. So this is an objection that the counterfactual theory can handle (albeit none too elegantly).

16. "Ensue," not "result:" since omissions are not causes, strictly speaking they have no results. See next Footnote.
17. Note well this word. What makes an action and an omission *correspond*? Not that they have the same consequences; since they are not causes, omissions have no consequences. However, in two parallel cases an action and an omission might be performed by two agents who want the same event to ensue either as a consequence (in the case of an action) or as something else that will foreseeably happen (in the case of an omission), e.g. the nephew's death. This means that, strictly speaking, an action and an omission can correspond only if the omission is intentional.
18. "If omissions as such show a lower degree of actionhood and responsibility, shouldn't it follow that there is *some* respect in which Callous Chris's behaviour is less bad than Interventionist Ian's?" No, because assessing Chris's behaviour as to whether it is an act or an omission is only part of what goes into assessing its overall moral character; and assessing its overall moral character is a holistic business—features that matter on their own can matter not at all "in the mix," and vice versa.
19. First stressed to me by Christopher Coope. He *did* make rather a meal of this point, but, as ever, helpfully.
20. "'Murder *by* omission'? Doesn't that phrase make the omission a *means to* murder, so that it must be part of the causal story?" No: the phrase tells us that callous Chris entertains a counterfactual of the form "If I don't do x, then y will follow." Such a counterfactual only implies a causal relation between Chris's omission of x and the ensuing y on the counterfactual model of causation. On the transmission model that I'm developing, the counterfactual only implies that Chris's omission so to speak *gets out of the way of* the causal relation that does his dirty work for him.
21. As noted by Jonathan Schaffer: see Howard-Snyder (2002, Note 25).
22. Thanks to Jo Wolff for this information.
23. Compare Mangan's classic statement: "A person may licitly perform an action that he foresees will produce a good effect and a bad effect provided that four conditions are [satisfied]: (1) that the action in itself from its very object be good or at least indifferent; (2) that the good effect and not the evil effect be intended; (3) that the good effect be not produced by means of the evil effect; (4) that there be a proportionately grave reason for permitting the evil effect" (Mangan, 1949, 43).

24. It would be very counterintuitive indeed to enounce an account of plans of action that made the bracketed words dispensable. Yet if we take seriously the suggestion that there could be an even partly extensional account of plans of action, it is hard to see why these words should *not* be dispensed with. This is another argument against extensional accounts.
25. McIntyre continues: "...and that his action is permissible even if he does intend to cause his own death as a means to save the others." Since this clause patently begs the question against PDE, I don't discuss it in the main text.
26. Of course, the agent would have to survive the explosion to be able to offer this report. But that is beside the main point at issue.
27. "Can be," not "is sure to be." As above, I am not saying that if actions of type A *can* be done with morally acceptable intentions, then any action of type A is acceptable.
28. So do "the numbers count," then? Yes they do, in this way: *provided no constraint is breached by so doing,* it is always permissible, and sometimes obligatory, to choose the comparable interest of the larger number over that of the smaller (Cp Taurek, 1977 who argues, I think implausibly, that choosing the interest of the larger number always breaches some constraint; against Taurek, see Parfit, 1978).

References

Austin, J.L. 1956. A plea for excuses. *Proceedings of the Aristotelian Society* 57:1–30.

Bennett, J. 1998. *The act itself.* Oxford: Oxford University Press.

Denyer, N. 1997. Is anything absolutely wrong? In *Human Lives: Critical Essays on Consequentialist Bioethics*, eds. D.S. Oderberg and J.A. Laing, 39–57. London: Macmillan.

Dowe, P. 2000. *Physical causation.* Cambridge, MA: Cambridge University Press.

Dowe, P. 2004. Causation and misconnections. *Philosophy of Science* 71:926–931.

Frankfurt, H. 1969. Alternate possibilities and moral responsibility. *Journal of Philosophy* 66: 829–839.

Harris, J. 1985. *The value of life.* London: Routledge.

Howard-Snyder, F. 2002. Doing and allowing harm. *Stanford Encyclopedia of Philosophy.* http://plato.stanford.edu/entries/doing-allowing/.

Kim, J. 1995. Causation. In *The Cambridge dictionary of philosophy, ed.* R. Audi, 110–112. Cambridge, MA: Cambridge University Press.

Mackie, J.L. 1974.*The cement of the Universe: A study of causation.* Oxford: Oxford University Press.

Mangan, J. 1949. An historical analysis of the principle of double effect. *Theological Studies* 10:41–61.

McIntyre, A. 2004. Doctrine of Double Effect. *Stanford Encyclopedia of Philosophy.* http://plato.stanford.edu/entries/double-effect/.

McMahan, J. 2002. *The ethics of killing: Problems at the margins of life.* Oxford: Oxford University Press.

Mill, J.S. 1872. *A System of Logic, 8th edition.* London: Online Library of Liberty. http://oll.libertyfund.org/index.php?option=com_staticxt&staticfile=show.php%3Ftitle=246&Itemid=27.

Parfit, D. 1978. Innumerate ethics. *Philosophy and Public Affairs* 7:285–301.

Rachels, J. 1975. Active and passive euthanasia. *New England Journal of Medicine* 2:8–80.

Salmon, W. 1997. Causality and explanation: A reply to two critiques. *Philosophy of Science* 64:461–477.
Salmon, W. 1998. *Causality and explanation*. Oxford: Oxford University Press.
Sartorio, C. 2005. A new asymmetry between actions and omissions. *Noûs* 39:460–482.
Shaw, J. 2006. Intention in ethics. *Canadian Journal of Philosophy* 36:187–223.
Sidgwick, H. 1874. *The methods of ethics*. Cambridge, MA: Cambridge University Press.
Thomson, J.J. 1985. Casual priority: A comment. *Noûs* 19:249–2.
Taurek, J. 1977. Should the numbers count? *Philosophy and Public Affairs* 6:293–316.

Part IV
Bioethics and the Natural Law: Challenges

Chapter 8
Global Bioethics and Natural Law

Ana S. Iltis

Proponents of global ethics frameworks seek to identify universally applicable, knowable and enforceable moral obligations and standards. Many of these are grounded in particular conceptions of human rights and seek to secure positive obligations to respect such rights. In bioethics there has been a special emphasis on global ethics. This is reflected in declarations and codes of ethics (World Medical Association, 1975, 1983, 1989, 1996, 2000, 2002, and 2004; UNESCO, 2005; Council of Europe, 1997) as well as in the bioethics literature, where some have challenged the plausibility of a global bioethics (Engelhardt, 2005; Cherry, 2002). Joseph Boyle makes important contributions to the literature on global ethics and global bioethics, nominating and defending natural law as a candidate framework for global ethics. He argues that natural law theorists have been doing for centuries what contemporary proponents of global ethics and global bioethics are trying to accomplish, namely identifying the moral framework applicable to and knowable by all persons in all times and places (Boyle, 2004). The conclusion that his interpretation of natural law is a plausible framework for global bioethics will be resisted by many contemporary bioethics scholars who propose global bioethics standards quite different from the new natural law that Boyle articulates with John Finnis and Germain Grisez (see Boyle, 2004, 2006; Finnis, 1980, 1983, 1991; Grisez, 1983, 1993, 1997; Finnis et al., 2001; Grisez et al., 1987). This chapter examines Boyle's contribution to the global bioethics literature and considers some of the difficulties associated with his attempt at framing a global bioethics.

8.1 Natural Law as a Framework for Global Bioethics

Global ethics emerges from a desire to have common moral standards that guide interactions among people in places near and far and, in some cases, from an interest in applying particular conceptions of human rights to biomedicine (Boyle, 2004, 2006). The UNESCO Universal Declaration on Bioethics and Human Rights

A.S. Iltis (✉)
Center for Bioethics, Health and Society, Wake Forest University, Winston-Salem, NC, USA
e-mail: iltisas@wfu.edu

C. Tollefsen (ed.), *Bioethics with Liberty and Justice*, Philosophy and Medicine 110, DOI 10.1007/978-90-481-9791-0_8, © Springer Science+Business Media B.V. 2011

articulates this goal: "it is necessary and timely for the international community to state universal principles that will provide a foundation for humanity's response to the ever-increasing dilemmas and controversies that science and technology present for humankind and for the environment" (UNESCO, 2005, preamble). Adopting such an ambitious goal leads to a very obvious set of questions: what are those principles, where do they come from, and how does one justify them? Natural law, Boyle argues, is the answer. Natural law refers "to the set of common moral principles, knowable in principle to all humans capable of thinking and choosing, which is the normative foundation for the precepts and virtues that comprise a moral life. These are the human goods and the principles ordering choice among them" (Boyle, 2006, 309).

According to Boyle, three characteristics make natural law a plausible candidate for a global bioethics framework. First is the accessibility of the precepts and principles of natural law to all persons through reason:

> some action-guiding thoughts . . . precepts or practical principles, are natural in the sense that they are not dependent for their validity on human decision, authority, or convention. Because of the independence of these factors, natural precepts and principles must be generally accessible to human reason; the critical reflection that is not dependent upon but potentially critical of any particular social enactment or practice is the work of common human reason (Boyle, 2004, 2).

Second is the universal applicability of the natural law. Natural law's principles and norms are "addressed to anyone at all facing a choice upon which the principles and norms bear" (Boyle, 2004, 2). As a universally knowable and applicable moral framework, natural law—if successful—provides precisely what proponents of global bioethics seek, namely a universal ethic that binds all persons.

Third, natural law responds satisfactorily to a common objection to the concept of global bioethics, namely that there is significant moral disagreement in the world rendering implausible a single account of morality knowable by and applicable to all persons. Natural law, Boyle argues, accommodates and explains diversity while articulating and accounting for shared moral space. Several features of natural law make this possible. One is natural law's account of a shared morality that is valuable and significant even though its norms may be shared for different reasons and there may be substantive disagreement about the nature of morality and the interpretation and application of norms:

> there are some moral convictions that are widely accepted across communities and across other barriers to moral agreement. ... These shared convictions are, to be sure, accepted by different people for various reasons, including purely local ones, and are commonly accepted without explicit reliance on moral theory. Among these moral convictions are some precepts addressed to anyone, for example, to respect the rights and liberties of other people. This universality cannot be dismissed as mere pretension, since such precepts have more than merely local standing, and this standing does not rely essentially on derivation from a controversial ethical theory (Boyle, 2004, 10).

At the same time, natural law theorists recognize that the shared moral convictions are not articulated independently of culture and time: individuals do not "cognitively stand above all moral traditions and communities" (Boyle, 2006, 310). As a result, "[n]atural law theorists ordinarily recognize that the intelligibilities that ground their moral beliefs can be articulated in more than one way, and have been given practical effect in a variety of social forms" (Boyle, 2006, 310). When individuals share moral common ground, they do not necessarily share it absolutely or for the same reasons. Boyle responds to objections to a universal ethic based on claims of obvious moral pluralism by arguing that underlying that pluralism there is common ground which is accounted for by natural law. Although such agreement appears coincidental, Boyle would argue that it is not. The apparent coincidence is the result of convergence through reason on principles that approximate the natural law. As discussed later in this chapter, such convergence may not be true or substantive. Even where there is agreement on an ethical principle, such as the impermissibility of murder, there may be radically different conceptions of which actions qualify as murder. Such differences may be the result of different understandings of who is innocent, when—if ever—it is permissible to execute a person found guilty of a crime, the circumstances under which guilt must be determined for an execution to be a just killing, and whether killings of certain beings such as fetuses constitute murder.

A second feature of natural law theory that enables it to explain the existence of moral diversity while maintaining that there is a universal morality knowable through reason alone is its account of moral reasoning and error:

> natural law theory . . . contains an account of moral error, of how individuals and communities can fail to know even fairly general and fundamental moral truths. Aquinas thought that people are often ignorant of, or mistaken about, the norms which should guide their choices. The norms about which ignorance and mistakes are possible include some fairly straightforward implications of moral principle. The cause for such ignorance is not limited to individual moral fault, which, of course, does explain much moral ignorance (Boyle, 2006, 316).

There is disagreement because some reasonable people are wrong; ignorance leads some people to moral error, which results in disagreements about norms. For Boyle, the fact of moral diversity does not serve as evidence to contradict the claim that a universal morality is knowable to all rational persons through reason alone. Boyle's claim is that differences in the interpretation of shared rational norms and human error explain the apparent moral disagreements we observe. Despite apparent moral diversity, the claim is that in fact there are shared moral norms that serve as the foundation for a global bioethics. This position seeks to discount the significant moral differences that characterize the human condition.

According to Boyle, the universal accessibility and applicability of the precepts and principles of natural law in addition to natural law's ability to respond to and account for evidence of moral pluralism that might otherwise be used to object to global bioethics render natural law the most plausible candidate in the search for a global bioethics framework.

8.2 The (Im)Plausibility of Natural Law as a Framework for Global Bioethics

In examining Boyle's vision of natural law as a framework for global bioethics, difficulties with his attempt to formulate a global bioethics become evident. I focus on four concerns here: reason as the source of moral knowledge; evidence of shared rational norms; the significance of vacuous norms; and use of state authority to enforce a universal morality.

8.2.1 Reason as the Source of Moral Knowledge

Boyle holds that the natural law is universally knowable to all persons through reason and that natural law is normative rather than merely descriptive or explanatory. This position is open to question, just as is any other claim that the content of morality is knowable to all persons through reason alone. No doubt many who share the desire for a universal bioethics and offer accounts of a global bioethics conceive of an ethic grounded in reason that is quite different from the one Boyle envisions. Boyle defends an interpretation or application of natural law known as the new natural law. According to new natural law theory, there are basic, intrinsically valuable human goods (see Boyle, 2006, 308; Grisez et al., 1987; Finnis et al., 2001; Finnis, 1980, 1983, 1991; Grisez, 1983, 1993 and 1997).[1] Although Finnis et al. offer different lists of such goods, they include among the basic goods against which one ought not intentionally act truth, peace, life, health, knowledge, play, friendship, religion, and aesthetic experience. Basic goods are goods the "pursuit of which seems of itself to promote persons and bring them together" (Finnis et al., 1987, 277). These goods are basic to what they term "human full-being." The basic goods give reasons for choosing and acting; they "provid[e] the reasons to consider some possibilities as choiceworthy opportunities" (Finnis et al., 1987, 277). Moreover, "[t]he reasons for choosing and acting, provided by basic human goods, require no prior reasons. The prospects of human fulfillment held out by peace, justice, knowledge, and so on, naturally arouse corresponding interests in human persons as potential agents" (Finnis et al., 1987, 278). According to the basic goods account, there is more than one morally worthy way of life and often more than one choice will be choiceworthy because "the possibilities of fulfillment are always unfolding, for there are several basic human goods, and endless ways of serving and sharing in them" (Finnis et al., 1987, 281). The moral life is on this account not monolithic. Moreover, it is not possible to fulfill maximally each basic good in any one life: "human beings, even when they work together, can only do so much. No one can undertake every project, or serve in every possible way. Nor can any community. Choices must be made" (Finnis et al., 1987, 281). Most choiceworthy choices require that we set aside other goods. Morally right choices are those made with the proper disposition, that is, with the aim of "acting for human goods and avoiding what is opposed to them" (Finnis et al., 1987, 283) and never intentionally acting against one or more basic goods. This first principle of morality leads to a second: "Do not do evil that good may come" (Finnis et al., 1987, 287).

The claims that these goods are intrinsically good, that one never may intentionally act against them, and that they are self-evident and can be known through reason, are open to question. It seems that we are supposed to recognize these goods as basic goods without a prior account of human nature and human fulfillment. Yet, to know that the basic goods identified in fact are intrinsically good because they are basic to human flourishing or human full-being, we need an account of human flourishing. Finnis et al. assert that "[g]oods like these are intrinsic aspects,—that is, real parts—of the integral fulfillment of persons" (Finnis et al., 1987, 277). To assert this requires knowledge of what it is for persons to be fulfilled, and it is precisely this account that is absent in Finnis, Boyle, and Grisez's work.

In addition to concerns over how we may come to know the basic goods, there is significant room for discussion of what it means to act against any of the basic goods. Consider contraception, the use of which Boyle holds is a violation of the basic good of life (Boyle, 1980; Grisez et al., 1988). Even if one were to accept the basic goods account and recognize that one ought never intentionally to act against the good of life, one might hold that contraception is not a violation of this good of life. Use of an abortifacient "contraceptive" might constitute such a violation, but use of a contraceptive per se—one that did not destroy a fertilized egg—would not violate life. Contraception does not involve the direct and intentional destruction of life, some would argue, and as such cannot violate the basic good of life. Some might go further and argue that it respects the good of life by allowing parents to bear only those children they can love, nourish and support.

A further concern is that other accounts of bioethics that also claim to rely on reason propose radically different moral content. Many of the actions contemporary natural law theorists understand as actions against the basic goods, such as the use of contraception which allegedly violates the good of life, are considered morally acceptable, and sometimes praiseworthy or obligatory, by other global bioethics proponents who similarly defend their account of morality based on reason and logic. Disagreement over abortion highlights this circumstance starkly. Boyle holds that direct abortion involves an intentional act against the good of life, which in all cases is prohibited (Boyle, 1987, 311). Others who similarly rely on reason or allegedly self-evident norms claim that a woman has a fundamental right to abortion and that this right must be protected. In fact, some hold that access to abortion is essential to the flourishing and well-being of women in society. In other words, what Boyle holds is self-evident or knowable through reason contradicts what other serious thinkers hold is knowable through reason or is self-evident.

Boyle responds to discrepancies in the moral knowledge supposedly secured through reason by arguing that some persons are wrong about what reason requires:

> Some philosophers' formulations of basic moral principles plainly contradict those of other philosophers. However, the merits of these different formulations can be, and often are, rationally debated. One purpose of that debate is to determine whether philosophers have

> introduced false assumptions into otherwise agreeable formulations of moral principle (Boyle, 2004, 5).

No amount of rational analysis and discussion may ever resolve such disagreements because attempts to access moral knowledge through reason alone necessarily require that we introduce additional assumptions, which cannot be proven, in order to justify the desired moral claims (Engelhardt, 1996; Engelhardt, 2000). For example, if we accept the norm that it is wrong to take the life of innocent persons, we must determine who counts as a person, when a person is innocent, and what counts as "taking a life." As Engelhardt has argued, one should not naively incorporate observations about the world into attempts to secure an understanding of morality by an appeal to rationality. There is no ideologically independent point from which to give an account of moral goods or their ordering:

> There is no standpoint from which canonically to balance moral claims or to determine which tradition or version of moral inquiry is more complete without begging the central question at issue: whose morality should govern (Engelhardt, 2000, 33)?

Or again:

> One cannot without prior guidance turn to nature to iscover criteria for appropriate human behavior or guidance in fashioning a bioethics. In a post-Darwinian world one must decide what adaptation, for which goals, in what environments should define biological success, health, and normality. In such circumstances, there are grounds to hold that there will not be a harmony among human moral dispositions, given the success that can be garnered by pursuing different goals in different environments. That is, in a post-Darwinian world that appreciates how environmental differences and multiform selective pressures produce biodiversity, a diversity of moral inclinations should not be unanticipated. One would even expect a biological basis for this diversity of moral sentiments, as well as what many would recognize as immoral inclinations, which in particular ecological niches will support inclusive fitness. There will then be alternative human moral types, but no ideal 'natural' type that can be accepted as morally normative for human sympathies and inclinations, not to mention actions (Engelhardt, 2000, 34).

And:

> To identify an account as normative, one must already possess a view of what is morally relevant in an appeal to facts, to nature, to human sensibilities, to human sympathies, or to human moral rationality. One must know how to sort moral information from noise. To discover a moral framework one can share with others, one must already share a basic background moral understanding. One must already possess a guiding moral sense, understanding, or narrative......To know how to rank values and moral principles, one must already have a moral or value standard in hand (Englehardt, 2000, 35).

Boyle's suggestion that we seek to identify false assumptions to eliminate error is insufficient; some assumptions may only be *deemed* true or false but not *shown to be* true or false. To demonstrate whether assumptions are true or false one must first establish rules of evidence, acceptable methods of inquiry, and rules of inference, all of which themselves are the objects of rational disagreement. Our moral disagreements about moral conclusions turn on disagreements concerning the veracity and legitimacy of the assumptions needed to identify norms. Without mechanisms to prove whether particular assumptions are true or false, we are left without final recourse through reason.

Boyle recommends that we demonstrate some epistemological modesty (Boyle, 2006, 314) when facing hard cases and that we avoid treating disagreement over hard cases as evidence of genuine and deep moral disagreement: "differences in moral judgment in difficult cases calling for careful casuistry to assess a host of conflicting moral considerations are also not taken by natural law theorists to be problematic, but to be expected given the complexity of these cases" (Boyle, 2004, 3). But his call for modesty and his warning that disagreement not be treated as evidence of pluralism are too weak, particularly in light of his failure to provide a rational ground to justify one ranking of the good and therefore one morality. This failure is not a failure of Boyle's per se but rather reflects the impossibility of justifying through reason alone one ranking of goods over all others. Reason alone is incapable of resolving such differences because certain assumptions are essential to making claims about morality. We may rationally debate the merit of different assumptions, but such rational debate will not necessarily lead us to knowledge for we always must rely on some basic metaphysical and epistemological assumptions and our post-Fall flawed humanity.[2] Hard cases may be hard in part because we as post-Fall humans lack the resources to resolve them.

Consider one of the central assumptions of natural law: nature has particular ends or goals which are knowable, in many cases self-evident (McInerny, 2004), and normative (see Iltis, 2004). While arguments can be articulated to defend this claim, we cannot prove the veracity of particular claims without begging the question. For example, should one consider moral norms to be realized in the character of each human or in a balance among different moral types (e.g., altruistic and egoistic), just as the balance among different blood types (e.g., "normal" and sickle-cell) may convey survival value? If the latter is the case, then it is "normal" and "morally appropriate" that some persons be selfish. These doubts concerning the character of natural moral norms are of a higher order than doubts about how to interpret nature's ends and what they require in terms of human action. Moreover, we must make assumptions about what we mean by nature and how we come to know nature if we are to rely on natural law (see Iltis, 2004). Given the role such assumptions play in shaping moral knowledge, it is unreasonable to expect agreement concerning the accuracy of such moral pronouncements. Even if there came a time when there was consensus on a moral claim assessed by all through reason, as fallen, imperfect beings, consensus among us does not guarantee accuracy.

Boyle responds to the claim that disagreements among rational people concerning what it is to have moral knowledge provides evidence that we cannot rely on reason to secure moral knowledge by arguing that those disagreements are not as radical as we might think: "defenders of natural law need not accept at face value the common formulation of ethical pluralism as involving moral disagreement at the most basic level" (Boyle, 2004, 4). Moreover, he holds, "it is not an incontrovertible fact that this diversity and disagreement goes all the way to the foundations of people's moral perspectives. The peaceful and cooperative interactions of diverse peoples surely suggests some common moral understandings and some agreements about common moral concerns" (Boyle, 2004, 4) and "many differences in custom and law among different societies are compatible with a common recognition of moral principle" (2004, 3). Reasonable people may disagree, for example, about whether a particular act constitutes a violation of one of the basic goods, but we all recognize that the basic goods are intrinsically good. Natural law accepts disagreements concerning the interpretation and application of norms with the expectation that over time, rational debate can help secure a broader or deeper shared understanding of morality. Given the diversity of fundamental assumptions that shape the character of what we call sound rational debate, this should not be anticipated. Extended disputes regarding abortion, homosexuality, and euthanasia have served to underscore the depth of the disagreements, not to resolve them. Moreover, as discussed below, Boyle's claim that we need not accept that disagreement is deep reflects wishful thinking and a failure to acknowledge that there are radically different conceptions of morality. Even if we can articulate some shared norms, conflict over some norms is so deep that any shared norm will be so thin as to make agreement on it morally irrelevant.

There is an additional objection to the natural law claim that we may secure moral knowledge through reason. Reason, according to some, is an inappropriate source of moral knowledge (Engelhardt, 2000; Bole, 2004). Knowing how appropriately to act requires knowing how appropriately to relate to a fully transcendent God. This in turn requires experiencing this personal God, not just understanding person-independent moral truths. Such persons do not argue merely that we cannot pull the rabbit (a canonical, content-full morality) out of the hat (rational philosophical argument) (Engelhardt, 2000, 89), but that moral knowledge is really about relating to the Person of God, not living in accord with supposed discursive rational moral truths. Moral knowledge is secured through experience and personal knowledge of God: "the project of morally knowing truly is in its core the project of turning away from self-love to love of God and one's neighbor, so as to experience God" (Engelhardt, 2000, 163).

Boyle would reject such views as misguided. He might appeal to some shared conclusions about morality between himself and persons who claim to reject reason as the source of moral knowledge as evidence of a shared morality. Or, he might argue that those who reject reason often in fact rely on reason to secure and explain moral knowledge. Finally, he holds that persons who reject reason cannot explain the moral agreement we observe in the world. The force of this response depends on there actually being moral agreement and such persons being unable to recognize or explain it. Someone like Engelhardt might argue that in fact what Boyle calls moral agreement is not moral agreement but rather prudential collaboration as in a free market with radically different moral commitments to exchange goods and services. Apparent moral agreement is explained as something other than real moral agreement.

We must recognize that there are not only radically different conceptions of morality with different views as to when, if ever, sexual acts (e.g., homosexual marriages) and homicide (voluntary euthanasia) are morally licit, but that there are fundamental disagreements regarding the nature of moral knowledge itself and of how one can acquire it. This should give great pause to persons who maintain that the path to moral knowledge is well-lit by reason and, as such, that we may all walk the same path and, eventually, reach the same conclusions about morality.

8.2.2 Evidence of Shared Rational Norms

A second difficulty with Boyle's approach to global bioethics concerns the extent to which the plausibility of his arguments depends on false claims about the existence of a shared understanding of morality. Moral pluralism is not as deep as it might appear, Boyle claims. Insofar as persons agree on some norms, even if for different reasons, there is nevertheless a universal morality that binds all persons. Boyle fails in eradicating the depth of pluralism by insisting that what some see as deep moral conflict is merely conflict over the interpretation of shared precepts and principles on which persons agree, so as to claim that that conflict signals a need for ongoing rational debate because such discussion may yield greater convergence.

Consider one of the norms Boyle takes to be shared: the obligation to respect the rights and liberties of individuals (Boyle, 2004, 10). This "shared norm" might be rejected by views of morality according to which the individual is not the basic moral unit. For example, some East Asian contributors to the bioethics literature have argued that the family is the basic focus of morality, so that moral obligations direct us to respect families, not individuals (Fan, 1997). For such persons, norms focusing on the individual will be understood as distorted, while a norm to respect the integrity of the family would be appreciated as justified. We must appreciate the far-reaching character of the implications of such radically different worldviews. The conflicts between the partisans of these moral understandings are not resolvable through rational argument, because persons are separated by basic assumptions regarding the character of morality.

Consider the implications for health care allocations of conflicting views as to whether property rights, equality rights, or the right to protection against morbidity and mortality risks is more basic. These disagreements are reflected, for example, in debates over whether providing tax-funded universal health care is essential to respecting the basic rights of individuals, or whether such policy violates basic rights and liberties. Even among those who hold that universal tax-funded health care must be provided, some hold only that a single tier system is legitimate while others believe a basic package should be provided for all with individuals left at liberty to purchase additional health care services. The significance of differences over the interpretation of moral norms is discussed below. Here the focus is on the depth of disagreement over norms.

Boyle argues against Engelhardt's articulation of secular morality by claiming that such a view is "incapable of explaining how certain, apparently universal norms, such as the requirement of promise-keeping, are morally justified" (Boyle, 2004, 6). The plausibility of Boyle's argument that apparently radical moral divisions are not as deep as they appear because there is a shared morality of promise-keeping depends on there in fact being such universally shared norms. The obligation to keep promises, he claims, is a shared norm. Only after a sophisticated analysis of what it means to make a promise and when one can be said to be keeping a promise could we justifiably claim that promise-keeping is a universal norm. For example, imagine that A, B, C, D, E and F all agree that one ought to keep one's promises. A promises B to repay a loan, but A understands a promise as a guarantee that an effort will be made, while B understands a promise as an assurance of payment. C promises to pay D in full for a loan D made to C at an interest rate D stipulated and to which C agreed, but C determines the rate of interest is unfair and that one must keep only promises that are fair. C begins making payments at a lower interest rate. E promises to repay F, but E makes this promise only because he must do so in order to get money from F, and E needs to secure the loan to help his brother avoid losing his home. E sees loyalty to family as not only permitting him to make a promise he does not intend to keep but as requiring it. The issues addressed by these scenarios are not hypothetical; they are at the heart of business and political struggles in many parts of the world.[3] Radically different understandings of the circumstances that excuse one from keeping a promise and of what counts as keeping a promise erode what it means to say that we all agree that we ought to keep our promises.

Even if we are justified in claiming that promise-keeping identifies a set of norms which share a family resemblance, we have stripped away all that is of interest and importance for the morality of appropriate action. Such stripped norms are relatively unimportant and uninteresting, as argued below, because once we strip a norm of content it loses moral significance. There may even be some who hold that promises are not "sacred" in themselves, that is, they do not accept the legitimacy of the norm that one ought to keep one's promises, but rather find the force of promises grounded in other moral considerations. They might, for example, maintain that promises hold in terms of being made to people with whom one has a close relationship and that it is this relationship that establishes promises as obligations that must be kept. If we do not share a background understanding of what it means to make a promise and when one is justified in making and breaking promises, then we do not mean the same thing by saying "I believe we ought to keep our promises." Agreement on vacuous norms does not facilitate mutual understanding or serve as a basis for moral interaction. Moreover, there are some who would disagree with the claim that we all ought to keep our promises because they hold that one ought not to make promises:

> Can people justifiably obligate themselves, in advance, to keeping their promises, in the face of other possible moral obligations? We argue that they cannot. Promising is, as a rule, immoral: it is either an advance declaration of the intention to do the immoral under knowable or unknowable contingencies, or else it is a deceptive, and thereby immoral, offer of assurance (Fox and DeMarco, 1993, 81).

The claim that the obligation to keep promises is universally recognized as a moral norm is open to question.

Boyle also claims that there are shared norms that specifically address bioethical issues and may provide a foothold for global bioethics. He cites the right to refuse treatment (Boyle, 2006, 330–332) and argues that the norm is shared despite different accounts of why individuals have a right to refuse treatment (331). One is the Western account of autonomy, but he claims that "modern autonomy is only one account of this norm and not the norm itself" (Boyle, 2006, 331). Boyle cites Cardoza in *Schloendorff v. Society of N.Y. Hospital* (1914) on consent to treatment and Pope Pius XII (1957/1958) on refusing treatment as evidence that the right is widely accepted and not an invention of bioethics in the 1970s (Boyle, 2006, 331). Boyle acknowledges that "[n]ot everyone and not every community accept this right and some try to keep it narrowly limited" (Boyle, 2006, 331), but suggests that the right to refuse treatment is the object of sufficient agreement to be part of global bioethics (Boyle, 2006, 332). Boyle assures us that the right is widely accepted, though the two examples he offers are Western. Moreover, a careful reading reveals that he initially proposes the right to refuse treatment as universally accepted and later acknowledges that it is not universally accepted but that it nevertheless is held by a sufficient number of people so as to include it in global bioethics. He offers no justification for claiming that the extent to which it is shared is sufficient and "good enough" for its being part of a normative global bioethics. Reliance on widely but not universally held norms is particularly problematic in establishing a global bioethics in light of efforts to enforce a universal bioethical morality, a concern discussed below.

Boyle cites a number of examples of norms he claims are universally shared. The universality of such norms is important for Boyle because the plausibility of his proposal that natural law serves as the framework for global bioethics turns in part on there being universally shared norms accessible to all people through reason, as he claims. In studying his examples, two concerns emerge. First, some of the norms he claims are universally shared are not shared because acceptance of such norms depends on metaphysical or epistemological assumptions that are not shared. Second, insofar as some norms are shared, they are shared in a morally uninteresting way because what is shared is terminology but not necessarily substance or content. There may be convergence on terms but norms are not shared in any deep sense because persons attribute different meanings to the same words.

An additional challenge for Boyle pertaining to the importance he gives to shared norms is that widely and even universally shared views regarding the truth are not, as a result of such a consensus, correct. There can be a serious disconnection between what people think is the case and what is the case, as with the once widely if not universally held belief that the sun revolves around the earth. That most people believe X is a moral norm that all ought to hold does not mean it is one. Even universal affirmation of a moral norm does not mean it is true.

8.2.3 The Significance of Vacuous Norms

A third concern with Boyle's discussion of global bioethics relates to a feature that he argues makes natural law a plausible framework for global bioethics. Boyle holds that by emphasizing the importance of shared norms that may be interpreted and understood differently, natural law accommodates and accounts for the moral diversity and disagreement we observe. What Boyle depicts as an advantage of natural law may require that we adopt such a thin view of what it means to share norms that convergence on norms is morally irrelevant. Where agreement refers to convergence on norms in the sense that persons agree on the terms but not on the meaning and interpretation of those terms, we should question whether there is morally relevant agreement, a problem identified above with regard to promise-keeping.

Boyle holds that even where there is disagreement about how to interpret a particular norm or the basis for holding such a norm, we should recognize that underlying the conflict is a shared understanding of morality and that it is this agreement that is important. He states that it is nearly universally held that persons have a right to refuse treatment (Boyle, 2006, 330–332), which generates the shared norm "that health care professionals refrain from treating competent adults without their consent" (Boyle, 2006, 330). Boyle recognizes the vagueness of terms like "competent adult" and "consent" but maintains that vagueness does not mean the norm is empty (Boyle, 2006, 330). Although it is true that "vague" does not necessarily mean "empty," Boyle discounts excessively the importance of defining terms. If we hold fundamentally different assumptions about what it means to be a competent adult, we do not share the norm to refrain from treating competent adults without their consent in a morally meaningful way. For example, in a culture in which one must be the male head of household to be considered a competent adult, the Western notion of individual consent for treatment or research participation will make no sense. To suggest that someone who holds the Western notion of individual consent and a person who believes the male head of household must give permission on behalf of others share a norm is absurd.

Where those who share a norm disagree radically about why the norm is a norm, or where they interpret and apply a norm in radically different ways, i.e., where those who share the norm disagree about what it requires and prohibits, it is questionable whether they truly share the norm. Is it fair to claim that those who allegedly share a norm but disagree about its foundation, its interpretation, and its application actually share a norm? Insofar as they do, the importance of such agreement is questionable. What is the moral significance of agreement on terms when that agreement does not translate to substantive shared understanding of action? Imagine Joe, Mike, and Sue all agree on the norm that lying is morally wrong. Joe holds this view because he is a Thomist and he understands that a lie is a falsehood told with the intention to deceive. Mike, on the other hand, is an act utilitarian. He agrees on the norm only because he believes that in most cases lying produces worse results than truth-telling and so he presumes the norm really is that "because the consequences usually are bad, it is morally wrong to lie." Implicit in his concurrence with the claim that lying is wrong is the understanding that the norm is situated within a broader understanding of maximizing the good. Sue agrees with Mike and Joe that lying is wrong, but she understands a lie to be a falsehood told to someone who has the right to the information in question. Is it true that they share a moral norm against lying? This state of affairs does not involve a "hard case." Three people with fundamentally different moral world views may nod and agree that they all hold a principle or norm in common, but the moral value of that agreement is negligible. They do not mean the same thing. It is comparable to saying that we agree that three dollars is a fair price for a loaf of bread, but if we do not specify whether we are talking about U.S., Canadian, or Australian dollars, our agreement may be meaningless (unless current exchange rates make the three currencies equivalent).

The concerns raised here are not hypothetical. Consider, for example, Angeles tan Alora's discussion of basic categories of morality, such as honesty. One might think that there is a shared understanding of what it means to be honest, even though we may disagree about when honesty is required or the degree of honesty required in a particular situation. She says,

> In the Philippines, honesty is understood in the context of personalism, which incorporates the concepts of *utang na loob* and *pakikisama*.[4] Determination of what is beneficial is viewed in relation to advantages to the person, or the person's group or family, rather than in terms of society as a whole. Contextualized in this way, honesty is an issue of personalism rather than legalism. It is a matter of loyalty rather than an issue of integrity. Honest action is evaluated not so much in light of its impact on the entire society as in terms of its function in maintaining important personal relationships (tan Alora, 2001, 61, note added).

As a result,

> An 'honest' person pays attention to personal relations within Filipino society. For instance, one does not steal from a person with whom one has a close relationship. . . ..Yet an 'honest' person may steal from an institution or the government. . ..Filipinos commonly use company supplies for private needs, submit padded bills, or even evade taxes (tan Alora, 2001, 61).

She adds that

> [s]uch a conception of honesty is manifested in medical education as well as in health care practice. For instance, cheating on classroom examinations or with regard to medical certificates is common. Rather than being condemned, such practices are rationalized as cultural norms, helping another person, or avoiding a greater evil (tan Alora, 2001, 61).

It is not that those involved think these are situations in which one may be dishonest but rather that they define honesty differently such that these behaviors are honest behaviors.

Terminological agreement absent significant substantive convergence allows individuals to make claims about a shared morality but such commonality is unhelpful in crafting policy, developing rules of engagement, and implementing codes of conduct, all of which are important in efforts to frame a global bioethics. Consider Boyle's claim that the obligation to respect the rights and liberties of individuals is a universally held norm. Despite my earlier objection to this claim, I grant it here for the sake of argument and assume it is a shared norm. If we agree that we ought to respect the rights and liberties of individuals but we have radically different understandings of what are the rights and liberties individuals possess, who possesses them, and what it is to respect them, such agreement is irrelevant for purposes of establishing a global bioethics.

As Mark J. Cherry notes, "Real disparity of belief threatens the possibility of reaching substantive consensus regarding liberty, autonomy, and basic human rights. When traditional constraints collide with the post-traditional individualistic search for self-understanding one engenders political controversy rather than moral resolution" (Cherry, 2004, 27). In commenting on Boyle's contentions that the National

Commission for the Protection of Human Subjects of Biomedical and Behavioral Research gives evidence of the possibility of identifying common moral norms in a pluralistic setting (see Boyle, 2004), Cherry notes that the consensus secured in the National Commission meetings was possible for two reasons. First, a number of controversial terms and concepts were left vague, such as what it means for subject selection to be equitable and what risks are considered daily life risks. Second, the parties involved did not represent the full spectrum of the moral diversity of our society or of the global community. Cherry notes that the history of the National Commission might have been quite different "had the Commission when discussing fetal research included the pope of Rome, an atheist feminist, a fundamentalist Baptist, a Maoist communist, a libertarian, and an advocate of unhindered choice in the matter of abortion" (Cherry, 2004, 27).

Boyle might counter that where there is disagreement, it either is a hard case or an instance of some participants introducing erroneous, false assumptions. He readily acknowledges that general norms will not always lead to shared conclusions about particular acts. For example, he says that as debates over the permissibility of physician assisted suicide, euthanasia, and the withdrawal or withholding of life-sustaining interventions demonstrate, even if we hold the general precept that we ought not to kill other human beings, in bioethics we face difficult questions about precisely what constitutes a killing (Boyle, 2006, 312). Boyle concludes that

> moral judgments in bioethical matters tend to fall along a continuum between the immediate implications of moral principles and the detailed casuistical judgments that require the analytical expertise of a trained casuist. To the extent such judgments fall closer to the end of this spectrum in which many diverse circumstances need to be considered, the prospects grow dimmer that we might have the capacity to articulate a body of bioethical norms that is at the same time circumstantially more nuanced than the Decalogue and useful for guiding bioethical decision making—that is, providing the guidance without complicated casuistical inquiry into circumstances (Boyle, 2006, 313).

This suggests that while the precepts of a common morality might be clear and universal, a shared understanding of how to interpret and apply them will not be a part of a global bioethics. Boyle is right to acknowledge these differences. What is puzzling is his interest in having us recognize the importance of shared norms when it is questionable whether they are shared in any morally relevant sense.

8.2.4 Use of State Authority to Enforce a Universal Morality

A fourth concern with Boyle's approach to global bioethics is that he holds that states may use their power and authority to enforce the supposedly universally knowable natural law that is applicable to all persons. Despite the absence of a global authority to regulate global bioethics matters, local leaders do have authority over such issues and may achieve something akin to global regulation by cooperating (Boyle, 2006, 312). Local leaders, for example, Boyle argues, have the authority to enforce positive welfare rights through taxation (Boyle, 2006, 312). In light of the uncertainty introduced by the imperfections of human reason, it is questionable whether it is appropriate to impose on all persons through the state a robust ethic that violates the moral convictions of some parties, e.g., the use of state force to provide what some believe are welfare rights by violating what others believe are property rights and in some cases forcing some persons to pay for what they believe are immoral goods or services. Any account of bioethics, global or otherwise, that seeks to use state authority and force to ensure compliance with an allegedly universal morality should be the object of moral suspicion. Consider Cherry's discussion of the use of state force to establish a particular moral vision:

> some argue that general beneficence together with respecting basic human rights and liberties requires everyone to participate in universal single-tier governmentally regulated health care, with tax-payer financed abortions for the impecunious. Those who disagree are to be coerced into participation through mandatory employer pay-roll tax deductions. Others argue that sustaining basic human rights and liberties requires understanding persons as free to refuse to be complicit in what one views as evil, with beneficence, including basic welfare needs, limited to acts of private charity (Cherry, 2004, 27–28).

And

> Taxation to support state-based health care is neither compatible with respecting real moral difference nor correctly described as peaceable social cooperation. Those who morally object to the provision of particular services are coercively compelled into participating in the purchase of such services (Cherry, 2004, 28).

The desire to use the state to enforce bioethics is not unique to Boyle. For example, the UNESCO Declaration directs states to adopt policies and laws that conform to and reflect the Universal Declaration: " The aims of this Declaration are: (a) to provide a universal framework of principles and procedures to guide States in the formulation of their legislation, policies or other instruments in the field of bioethics..." (UNESCO, 2005, Article 2a).

H. Tristram Engelhardt, Jr. has addressed the permissibility of using force in the face of pluralism and diversity. Engelhardt argues that insofar as we seek peaceably to cooperate with moral strangers, persons with whom we do not share a content-full morality and who will not come either through conversion or sound rational argument to share our morality, we must recognize and respect the principle of permission. That is,

> Authority for actions involving others in a secular pluralist society is derived from their permission. As a consequence,
> i. Without such permission or consent there is no authority.
> ii. Actions against such authority are blameworthy in the sense of placing a violator outside the moral community in general, and making licit (but not obligatory) retaliatory, defensive, or punitive force (Engelhardt, 1996, 122).

Engelhardt's analysis raises critical doubts about the extent to which individuals or states may use force in the effort to create and sustain a global bioethics.

8.3 Conclusion

Joseph Boyle makes important contributions to the global bioethics literature. First, he argues that insofar as global bioethics proponents seek a universally knowable and applicable moral framework that can account for present-day moral pluralism, they should take natural law seriously. Those who propose a global bioethics framework grounded in something else should be prepared to defend the superiority of their account to natural law, which has been the subject of rigorous analysis and refinement for centuries. They also must be prepared, as must any natural account, to respond convincingly to limits global bioethics frameworks face, including those discussed here.

Notes

1. For an insightful discussion of the basic goods and new natural law, see DuBois (2006).
2. After the Fall, man's will and intellect are unreliable and nature is broken:

   > It is not simply that after the Fall man is subject to the passions and weakness of will, and is therefore unable to carry through by himself the resolution to do the good. It is also the case that his intellect, his noetic capacities no longer without struggle allow him discernment of good and evil undistored by desire. He also no longer possesses spiritual knowledge. The impact of the Fall is not so much on man's will as often supposed in the West, but upon his intellect, his noetic capacity for non-discursive knowledge
   >
   > Nature as creation is also now broken. In many ways, what is now 'natural' is that which is improper for, and turned against, humans. The nature of man himself as well as the physical and biological nature that surrounds him is deaf to human purposes, if not hostile to them. The curse that comes to Adam with the Fall includes the world's malignancy" (Engelhardt, 2000, 174).

3. See Hisrick et al. (2003) for a discussion of business ethics in general and different understandings of the role of promises in business relationships in Russia, Slovenia, Turkey, and the United States. See Moon and Wooliams (2000) for a discussion of differences among cultures regarding issues such as rights and loyalty.
4. *Utang na loob* refers to "a concept of reciprocity or debt of gratitude or honor that imposes corresponding obligations and behavioral expectations. It imposes on the recipient of a good act or deed to behave generously toward the benefactor for as long as they both live" (tan Alora, 2001, 12). *Pakikisama* refers to "the ethos of loyalty, unity, peace, and cooperation demanded by the strong familial community among Filipinos" (tan Alora, 2001, 11).

References

tan Alora, A. 2001. Honesty, loyalty, and cheating. In tan Alora and J. Lumitao (Eds.) *beyond a Western bioethics: Voices from the developing world* 60–66. Washington, DC: Georgetown University Press.

Bole, T.J. 2004. The perversity of Thomistic Natural Law theory: Reflections on Lustig's criticisms. In *Natural Law and the possibility of a global bioethics* ed. M.J. Cherry, 141–147. Dordrecht, The Netherlands: Kluwer academic.

Boyle, J. 1980. Contraception and Natural Family Planning, *International Review of Natural Family Planning* 4:309–315.
Boyle, J. 2004. Natural Law and global ethics. *Natural Law and the possibility of a global ethics* ed. M.J. Cherry, 1–15. Dordrecht, The Netherlands: Kluwer academic.
Boyle, J. 2006. The bioethics of global biomedicine: A Natural Law reflection. In *Global bioethics: The collapse of consensus* ed. H.T. Engelhardt, Jr, 300–334. Salem, MA: M&M Scrivener Press.
Cherry, M.J. 2002. The search for a global bioethics: Fraudulent claims and false promises. *Journal of Medicine and Philosophy* 27:683–698.
Cherry, M.J. 2004. Natural Law and moral pluralism. In *Natural Law and the possibility of a global ethics* ed. M.J. Cherry, 17–38. Dordrecht, The Netherlands: Kluwer academic.
Council of Europe 1997. Convention for the Protection of Human Rights and dignity of the human being with regard to the application of biology and medicine: Convention on Human Rights and biomedicine. Oviedo. http://conventions.coe.int/treaty/en/treaties/html/164.htm (Accessed July 5, 2007)
DuBois J.M. 2006. How much guidance can a secular, natural law ethic offer? In *The death of metaphysics; the death of culture: epistemology, metaphysics, and morality*ed. M.J. Cherry, 185–197. Dordrecht, The Netherlands: Springer.
Engelhardt, H.T., Jr. 1996. *The foundations of bioethics, 2nd edition*. New York: Oxford University Press.
Engelhardt, H.T., Jr. 2000. *The foundations of Christian bioethics*. Lisse, the Netherlands: Swets and Zeitlinger.
Engelhardt, H.T., Jr. 2005. Critical care: Why there is no global bioethics. *Current Opinion in Critical Care* 11:605–609.
Engelhardt, H.T., Jr. 2006. The search for a global morality: Bioethics, the culture wars and moral diversity. In *Global bioethics: The collapse of consensus* ed. H. T. Engelhardt, Jr. 18–49. Salem, MA: M&M Scrivener Press.
Fan, R. 1997. Self-determination vs. Family-determination: Two incommensurable principles of autonomy. *Bioethics* 11:309–322.
Finnis, J. 1980. *Natural Law and Natural Rights*. Oxford: Clarendon Press.
Finnis, J. 1983. *Fundamentals of ethics*. Washington, DC: Georgetown University Press.
Finnis, J. 1991. *Moral absolutes: Tradition, revision, and truth*. Washington, DC: Catholic University Press.
Finnis, J., J. Boyle, and G. Grisez. 2001. *Nuclear deterrence, morality and realism*. New York: Oxford University Press.
Grisez, G. 1983, 1993, 1997. *The way of the Lord Jesus: Volumes 1–3*. Quincy, IL: Franciscan Press.
Grisez, G., J. Boyle, J. Finnis, and W.E. May. 1988. Every marital act ought to be open to new life: Toward a clearer understanding, *The Thomist* 52:365–426.
Grisez, G., J. Finnis, and J. Boyle. 1987. Practical principles, moral truth, and ultimate ends. *The American Journal of Jurisprudence* 32:99–151.
Fox, R.M. and J. DeMarco. 1993. The immorality of promising. *Journal of Value Inquiry* 27:81–84.
Hisrick, R., Bucar, B., and S. Oztark. 2003. A cross-cultural comparison of business ethics: Cases of Russia, Slovenia, Turkey, and United States. *Cross Cultural Management* 10: 3–28.
Iltis, A. 2004. An assessment of the requirements of the study of Natural Law. In *Natural Law and the possibility of a global ethics* ed. M.J. Cherry, 115–122. Dordrecht, The Netherlands: Kluwer academic.
McInerny, D. 2004. Natural Law and conflict. In *Natural Law and the possibility of a global bioethics* ed. M.J. Cherry, 89–100. Dordrecth, The Netherlands: Kluwer academic.
Moon, C. and P. Woolliams. 2000. Managing cross cultural business ethics. *Journal of Business Ethics* 27:105–115.

Pius XII 1958. The prolongation of life: Allocution to the international congress of Anesthesiologist – November 24, 1957. *The Pope Speaks* 4:395–396.
Schloendorff v. Society of New York Hospital 211 N.Y. 125, 105 N.E. 92 (1914).
UNESCO 2005. *Universal declaration on bioethics and Human Rights*. http://portal.unesco.org/en/ev.php-URL_ID=31058&URL_;DO=DO_TOPIC&_SECTION=201.html (Accessed July 5, 2007).
World Medical Association 1975, 1983, 1989, 1996, 2000, 2002, and 2004. *Declaration of Helsinki*. http://www.wma.net/e/policy/b3.htm. (Accessed July 5, 2007).

Chapter 9
Guided Autonomy and Good Friend Physicians

Janet Smith

The professional discipline of bioethics for some time now has seemed to be concerned primarily with two major questions.[1] One is how various ethical theories such as utilitarianism, deontology, natural law ethics, principlism, feminist care ethics or casuistry (and perhaps other systems as well) would assess a certain action, e.g., assisted suicide or cloning.

Yet few patients or physicians are strict advocates of any of the usual ethical positions proposed; individuals rarely identify themselves as members of any of the groups listed above. Knowing how various philosophical systems may judge a particular action helps few health care professionals[2] know what to do in their day-to-day lives. Most people, physicians and patients, it seems to me, are either religious believers, in which case their faith is a very strong influence on their ethical thinking, or they are agnostic about any kind of ultimate commitments, and they make their decisions either in accord with what their culture allows or in accord with some not entirely coherent set of personal beliefs. The number of individuals who consciously and deliberately make their moral decisions according to reflectively established ethical commitments (as opposed to religious commitments) is possibly nearly negligible. Moreover, the focus on various ethical positions sometimes seems to obscure the reality that ethics is ultimately a practical science that is ordained to ethical choices in the particular; it is not ultimately ordained to knowing and justifying universal claims and commitments. It is my view that bioethics pays too much attention to sharpening philosophical justifications than to the fact that many patients and physicians need help in making very particular decisions and are rarely helped in those decisions by philosophical clarifications.

A second major concern of contemporary bioethics is ensuring that a patient makes as autonomous a choice as possible; thus a great deal of attention is given to informed consent, advance directives, proxies, etc.

One would think from what one often reads in bioethical texts that the role of the physician in a difficult medical/ethical decision is to present all the relevant medical information to the patient, to inform the patient of all legal options open to

J. Smith (✉)
Sacred Heart Major Seminary, Detroit, MI, USA
e-mail: profjanetsmith@comcast.net

C. Tollefsen (ed.), *Bioethics with Liberty and Justice*, Philosophy and Medicine 110,
DOI 10.1007/978-90-481-9791-0_9, © Springer Science+Business Media B.V. 2011

him/her, and to remain as uninvolved in the decision-making as possible so that the patient may make a fully autonomous choice. To be more involved would be to be guilty of paternalism. Yet often patients are not prepared to make the autonomous choices bioethicists want to ensure for them. Patients are not nearly as desirous of being autonomous as bioethicists might think or want them to be. They may in fact perhaps want and need considerable assistance in making decisions.[3] Even those who profess to be religious believers, even of denominations with dogmatic teachings, are often uncertain about what choices available to them are in accord with the principles of their faith.

Perhaps we pay little attention to the particular decision because we are so devoted to respect for autonomy. The position that respect for autonomy should be the premier value in bioethics is problematic, in my view.[4] Our devotion to autonomy is to some extent rooted in an anthropology that understands free will to be a defining feature of the human person. Autonomy, however, seems the premier value not so much because of anthropological commitments but largely because of epistemological commitments, because of a general skepticism; we simply do not believe that certain knowledge is possible, especially about moral matters. Even many of those who profess a personal certitude about some moral issues believe that in a pluralistic culture such as ours, we must not impose our views on others. Our age is one that wishes to exercise tolerance of the views and choices of others as much as possible.

We do, of course, put some limits on the exercise of autonomy both within the culture and within the medical profession. Those limits are clearest when our actions harm others or cause social disorder; we do not permit killing and rape and theft, for instance. Yet we also put fairly intrusive restrictions on individuals for their own sake; we require seatbelts for those operating and occupying motor vehicles and helmets for motorcyclists. Medical procedures and drugs are carefully regulated. Few bioethicists or physicians are radical autonomists who support the "right" of the individual to do whatever he/she chooses; few consider physicians and surgeons to be mere hired technicians who do the bidding of their "customers." Most are what I would call moderate autonomists, those who think the community can place some restrictions on choices; moderate autonomist physicians will perform procedures that are legal and refrain from illegal ones. I am advocating a position I call "guided autonomy;" one holding this position would hold autonomy in high regard but also believe that there are times when it is morally right and obligatory to try to influence the choices of others, even when their choices are legal.[5] I will lay out the principles justifying guided autonomy below.

Bioethics has not, of course, been totally oblivious to the particular. In fact, as a subdiscipline of "applied ethics" bioethics regularly makes use of case studies to illustrate the application of principles and also the conflicts that easily arise at the bedside.[6] There has been abundant awareness that the medical facts of a particular situation, the moral views of the participants in the decisions (e.g., the patient, the health care team, the law, the relatives and friends of the patients), the resources of the patient, hospital, and community are among the many factors that come into play in particular bioethical decisions. What has not been done, to my knowledge, is to attempt to catalogue what types of influence, what methods of persuasion, manipulation, or coercion[7] (I will clarify these terms more fully below) are morally permissible for those who would guide a patient to a certain decision, and when it would be morally permissible to employ various methods. Here I hope to provoke bioethicists to consider these questions and to provide a modest beginning to that consideration.

The question I am going to explore here is the question of what kind of techniques a health care professional may morally employ to attempt to influence the medical/moral decisions of patients. I am not speaking of clear conscience cases where a physician could rightly find some way to withdraw from the case but of situations where the physician believes that a continued relationship with the patient is warranted and that an effort to persuade or guide the patient is part of his or her moral obligation. For instance, what should a physician do when a patient exhibits a kind of death wish and refuses treatment that has a high probability to cure or at least make manageable a lethal or somewhat disabling disease? Or even when a patient is reluctant to initiate a change in lifestyle or to embark upon a treatment that may not be directed towards an immediately life-threatening condition but may be much advised to prevent such conditions from arising. How much and what kind of influence is it moral for a physician to exercise on a patient to try to convince him or her to do what the physician believes to be the sensible and moral thing to do? I believe the focus of bioethics on explaining the fundamental principles that govern the ethical evaluations of various systems and on ensuring respect for autonomy have led bioethics to neglect helping health care professionals make a kind of moral decision that arises regularly in their practices.

The discipline of bioethics and our legal system[8] (and the religious tradition of which I am a member, Catholicism) frequently invokes the principle that patients have the right to refuse medical treatment and often speaks of "any and all" medical treatment.[9] To force medical treatment on a patient is considered a violation of the body, equivalent to assault, and also a violation of a person's rightful sovereignty over his/her body and choices, increasingly the "right to refuse medical treatment" is *spoken* of as an absolute right. Nonetheless it is definitely not *treated* as an absolute right; indeed, with the exception of a few notable cases, the courts have generally refused to penalize physicians who have disregarded patient's directives by forcing them to undergo life-saving measures (see Malloy, 1998, 11). Moreover, physicians routinely use a variety of methods to attempt to move patients to accept treatment they may oppose. Here I am asking the bioethics and health care professions to reconsider their verbal commitment to non-intrusiveness, especially in light of what seems to me widespread violation of that commitment in practice.

The fact is that health care professionals regularly face patients who are reluctant to undergo recommended treatment and they labor to find some way to lead patients to adopt the recommend treatment. To use the terminology of the day, this could be characterized as a situation where the principle of beneficence conflicts with the principle of respect for autonomy. Yet, in the bioethical literature this situation is not treated as a day-to-day problem but as a rather extraordinary problem with the dominating example being the Jehovah Witness who refuses to have a blood transfusion (because of various alternate treatments that are available, the problem is not nearly so urgent now as decades ago). Several of the elements upon which I hope to focus do come into play in the Jehovah Witness example since it involves a routine practice that medical science holds to be an objectively right and good treatment for many medical conditions and a patient who refuses that treatment for reasons not widely shared. But while this is an instance of the kind of case I wish to consider (and I will return to it), it has a component that I believe is somewhat unrepresentative of cases that interest me and that is the element of a deeply held religious conviction. Here, I am primarily interested in cases where the request for treatment or the refusal of treatment stems from much less lofty causes, such as ignorance, stubbornness, quirkiness, or depression. Sometimes these situations are treated, at least theoretically, as though individuals were being guided by deeply held religious convictions and that their right to act in accord with their own consciences ought not to be violated, when in fact, sometimes the patients are acting more like intransigent children refusing to listen to good counsel than fully mature adults making carefully considered autonomous judgments.

Now, I believe that nearly every health care professional, no matter how relativistic in his/her philosophical commitments, and no matter how committed to the principle of respect for autonomy, routinely utilizes a bag of tricks for persuasion, manipulation and coercion not unlike those of a parent dealing with a willful child (below I shall identify more explicitly what some of those techniques are). I am going to focus on the appropriate response for those who have a set of philosophical and religious commitments that in fact necessitate that the physician attempt to influence those who are considering making choices the physician believes to be immoral. Most Christians hold these beliefs; stated briefly, they are:

1. Human freedom is of the essence of the person.
2. Human dignity resides in choosing what is good freely; virtue, which is necessary for human happiness, can only be acquired through action flowing from free choice.
3. Choosing what is good freely is the highest value, not just acting freely;
4. Individual autonomy is not the highest value, especially an autonomy to be exercised free of influence;
5. Doing what is immoral is a serious matter—it may endanger the immortal soul of the agent and those assisting the agent and also have seriously bad repercussions for society. (These first two commitments are very characteristic of religious believers, of course, but such pagan philosophers as Plato and Aristotle also held them);
6. We have a moral obligation to help others make good moral decisions; the more involved our lives are with theirs, the greater the obligation;
7. There is an objective reality that yields some moral absolutes;
8. Ethics is not ultimately about universal, objective, moral absolutes, but is about concrete particular actions made by a concrete particular unique individual;
9. Although there is an objective reality yielding an objective morality to which we all must submit, in most concrete situations, there is a legitimate range of

ethical choices in which an individual's own existential situation and sometimes even preferences are of considerable importance in determining which is the ethical choice for him or her in a particular situation. That is, there is a degree of subjectivity and relativity in ethics;

10. That the individual conscience is of extreme importance; that in many matters, we need to honor the conscientious choices of others even when we have moral certitude that the individual is seriously wrong;
11. That in some choices the consciences of many are involved; conflicts will sometimes be inevitable and not resolvable. Sometimes it will be the case that the person with the most might, will be right to exert that might.

I am not going to directly justify these commitments, though some of what I write below would serve as a basis for a justification; I want rather to focus not on the reasons for these commitments, but how those holding these commitments are to function in practical situations.

This paper is going to have the curious shape of arguing on the one hand that physicians, at least those holding the above philosophical commitments, have both a moral obligation to try to influence the moral decisions of their patients, sometimes to the point of getting court orders to force the patient to undergo a treatment, and on the other hand, that they also have a moral obligation to help their patients make their own decisions, and sometimes must not intervene in the wrong moral decision, even when the decision leads to the patient's unnecessary death.

Those who hold the above positions are not really paternalists; they do not think they know what is best for a patient because they are the all-wise and knowing physician. Rather they are concerned to help a patient exercise his autonomy well, that is, in accord with the truth. Yes, it seems arrogant to act as though one knows the truth and to be prepared to guide others to follow the truth, but isn't it an act of friendship rather than arrogance to try to help others make good decisions? All the laws we have that make many drugs inaccessible without a physician's approval are paternalistic and coercive in some respects; among other premises, they are based on the premise that there are some bad decisions that people ought not to be allowed to make. Good friend physicians respect immensely the autonomy of their patients but also have a commitment to objective truth; thus they will try to guide their patients to exercise their autonomy to act in accord with the truth.

The "guided autonomy" dilemma—the dilemma of those who respect personal freedom and autonomy but also believe in objective truth and who want to act as friends to others—is not, of course, a dilemma distinctive of bioethics; it permeates all human relationships. In fact, let me give an example that haunts many a parent.

Consider the parent who believes his/her adult child to be prepared to marry someone highly unsuitable, someone the parent is convinced the adult child would regret marrying. Certainly, adults are free to marry whom they choose, barring legal impediments. Nonetheless, many factors other than the legal impediments might motivate a parent to attempt to prevent a marriage from happening. Now, the extent of the harm and regret that an adult child might experience would be a major factor.

Suppose one saw signs that the fiancé would be physically and psychologically abusive, was a homosexual, or would return to a life a crime or drugs? In this instance, a parent might attempt to exercise an enormous amount of influence. A parent might offer to pay for a trip abroad for the adult child to put some distance between the lovers; a parent might enlist the assistance of a friend, family member, or respected individual who might influence an adult child. If there were some outstanding arrest warrant against the fiancé parents might alert the police to the whereabouts of the fiancé. If a parent knew someone with extremely damaging information about a fiancé's past that would alarm one's child, the parent might go to great lengths to get that information. If a parent could track down a former and still interested and still appealing suitor, one perhaps in the past thought not altogether desirable but much more desirable than the current fiancé, a parent might get that former suitor into the picture. All the while the parents might completely accept the principle that adults have the right to marry whom they choose but would still feel themselves obligated to do everything they could to prevent a marriage.

The physician/patient relationship is clearly not as close as that of parent/child and undoubtedly my example is making ice-water flow in the veins of those who denounce paternalism, but when a profession deals regularly with issues of life and death, clinical detachment is also not the right response. If one is not right to exercise the control that a parent would with a minor child, it may nonetheless be right to attempt to have at least the influence that a good friend would have. For those of us who still retain vague memories of a family physician who was nearly part of the family, letting go of the idea of the physician as one who cares deeply about one in every respect is difficult.

So the task here is to try to articulate some principles that would enable good friend physicians both to respect the autonomy of their patients but also to respect objective truth and morality. I am trying (and inviting others to help me) to articulate some principles or at least bring into focus some guidelines for knowing when and how to influence a patient's decision and knowing when and how to recognize when there is a range of decisions that could be morally legitimate for a particular patient and when a particular patient should be free to make his or her decision unencumbered.

Let me for the moment draw upon the thought of Aristotle. I do so because he regularly compared the training of others in virtue and the living of the ethical life to the practice of medicine. He regularly notes that the physician has a fixed goal: health, just as the ethical person has a fixed goal: happiness, or activity in accord with virtue. He notes that both operate by attempting to know what is universally true of all human beings and also what is true of this particular person. For instance, we find this passage in the *Metaphysics*:

> …. experience is knowledge of individuals, art of universals, and actions and productions are all concerned with the individual; for the physician does not cure a man, except in an incidental way, but Callias or Socrates or some other called by some such individual name, who happens to be a man. If, then, a man has theory without experience, and knows the universal but does not know the individual included in this, he will often fail to cure; for it is the individual that is to be cured (Ross, 1979, 981a15ff).

Aristotle makes many comparisons between ethics and medicine because both are practical sciences: both seek knowledge for the sake of action.

> For the end of theoretical knowledge is truth, while that of practical knowledge is action (for even if they consider how things are, practical men do not study what is eternal but what stands in some relation at some time) (Ross, 1979, 993b20).

That is, truth is used by a practical science not as an end in itself but as a means to something else or for something temporary which itself is an end in itself; for example, one seeks ethical truth for the sake of virtuous activity; one knows about drugs for the purpose of healing.

The virtue that enables one to make the right particular choice, that is, to apply the universal correctly, is the virtue of prudence. (In reading the passage below, keep in mind that Aristotle thought nutrition was an expertise of the physician):

> Now prudence is not limited to what is universal but must know also the particulars; for it is practical, and *action* is concerned with particulars. And it is in view of this that some other men, without universal knowledge but with experience in other things, are more practical than those who have universal knowledge only; for if a man knew universally that light meats are digestible and healthy but did not know what kinds of meats are light, he would not produce health, but a man who knows that chicken is light and healthy is more likely to produce health. Now prudence is concerned with *actions*; so we should have both kinds of knowledge, or else the latter rather than the former, which is universal (Apostle, 1984, 1141b15-23).

That the *Nicomachean Ethics* is ultimately directed towards the particular is not often noticed. There are some significant passages that give general advice about how to make particular choices, for instance, certainly *Nicomachean Ethics* 1.13; much of the discussion about the mean, and the books on friendship. One passage makes a wonderful parallel between ethics and medicine:

> . . .private education is superior to group education, as in the case of medical treatment; for though in general rest and abstinence from food are beneficial to a man with fever to a particular man perhaps this is not so, and perhaps a boxing instructor does not prescribe the same mode of boxing to all his students. So it would seem that greater accuracy in detail is attained if each person is attended to privately, for in this way he is more likely to receive what suits him. But a physician or an athletic instructor or any expert can best attend to an individual if he knows universally, i.e., that such and such is the case of all men or for all men of a certain kind, for scientific knowledge is predicated of what is common and is universal. Now perhaps nothing prevents even an unscientific man from attending well to some one thing, e.g., to a certain man, if he has accurately observed through experience what happens to him, like some people who seem to be their own best doctors but are unable to help anybody else. But if one wishes to become an artist or a scientist, it would seem that he should none the less proceed to the universal and know it as far as possible, for we have stated that sciences are concerned with universals (Apostle, 1984, 1180b8-23).

The emphasis in this passage is on knowledge of the universal but clearly the particular is stressed as well. Earlier in this chapter Aristotle had given instruction on how best to teach different sorts of individuals to be virtuous; one kind of instruction is suitable for the noble, another for those who are motivated by fear of punishment, and yet another for those who are incurable. Sometimes inspiring stories are sufficient, sometimes paternal instruction is best; sometimes the force of law is needed.

Aristotle finds the following parallel between ethical and medical reasoning. Medical reasoning often proceeds in the following fashion: X is the healthy state of a certain organ (e.g., a healthy immune system fights disease); Y (or Ys) causes a certain disease or causes a bodily part to malfunction (e.g., HIV causes AIDS which damages the immune system); and Z (or Zs) treats or cures a certain disease or corrects a malfunction (e.g. antiretroviral drugs treat HIV). Virtue ethics reasons in a similar fashion: A is a virtue that all should strive to achieve (e.g., justice); and B is a kind of action that is always incompatible with the acquisition or maintenance of a certain virtue (e.g., paying a unjust wage); C is the kind of action that will advance or restore a virtue (e.g., paying a just wage). Both are nonetheless ultimately concerned with the particular; is X^1 level of health is the best we can hope for Joe the patient with the HIV virus (e.g., no onset of AIDS for perhaps 15 years); Y^1 is the reason for this condition in this patient (sexual promiscuity); Z^1 is the best treatment (e.g., a certain antiretroviral drug). In the case of character building for Jane, the manager of a McDonald's franchise seeking to be a just person, she would want to pay a fair wage to her employees—more for those who have worked many years, less for the neophyte.

Aristotle thought both physical and ethical goodness were in some sense the same for everyone; he thought everyone naturally desired health and that what health is can be determined by understanding the nature of the human body; it is designed to fulfill its natural functions properly and if those functions are operating in such a way as to make their natural contribution to the whole, they would be considered healthy. Thus, the ancients considered health care to be a profession with some considerable inherent dimension of paternalism certainly insofar as the physician knew more medicine than the patient, but also because the physician strove to do what was best for the patient—the health of the patient was his or her concern, not fulfilling the various wishes of a patient for which the physician had the skills to deliver.

There would, however, be relative senses of health for particular individuals; what one can expect about how the body of an 18 year old male would function is different from how the body of an 80 year old woman would function.[10] Exercise, nutrition, health care would differ accordingly. Aristotle thought moral goodness was similarly universal; all should strive to possess the virtues, though what would constitute virtuous behavior for those in different states of life would vary.

So, first let me note, that even those who believe there are objective moral norms and objectively right or superior medical treatments will demonstrate great flexibility and variety in ethical and medical judgments because of the different factors in individuals' lives. For a 35 year old mother of small children with breast cancer, a double mastectomy followed by chemotherapy and radiation may be the obvious medical choice and even the obvious moral choice; whereas a 90 year old woman might have both good medical and moral reasons for refusing the same treatment. But some 35 year old woman might also be within her moral rights to refuse the treatment (e.g., someone who had multiple other medical problems and who might not survive the operation or treatment) and some 90 year olds might be under an obligation to pursue it (e.g., whose presence was of extreme importance to a child).

A remarkable story in the *Atlantic Monthly* reported on patients who wanted to have a healthy limb amputated, some for kinky sexual reasons, some for identity reasons—they identified themselves with amputees (Elliott, 2000). Besides this particular desire, they exhibited no manifest signs of insanity. Evidently most would be able to pass a psychological test demonstrating competency to make their own decisions. A question raised by the article is this: why should a doctor who believes that patients' bodies are their own to do with what they will and that patients have a right to fully autonomous choices, refuse to perform such surgery? And then there is "Lizard man,"[11] the philosophy graduate student who has been undergoing a series of plastic surgeries to transform his appearance into that of a lizard—he is doing an experiment to determine how people respond to a lizard who acts like a human being. He is doing with his body what he wants and some physician(s) is willing to assist him. Very few physicians would amputate healthy limbs or do these surgeries, yet, I suspect it would not be their respect for autonomy, but their love of money that would guide their behavior. Those who hold to Aristotle's understanding of health as a knowable standard, the guiding autonomists or good friend physicians, and those who view the physician as one who works to help his/her patients attain the standard of health appropriate to them, would have no problem justifying their refusal to perform the above procedures; those who hold to a respect for autonomy would have some problem.

But simple refusal to perform a procedure is sometimes not right. What to do with a patient who refuses to pursue a treatment plan that a physician believes would be of great health benefit? Let me first ask this question in respect to a situation where the patient's resistance to the physician's recommendations is not of a moral nature. Patients may be, for instance, reluctant to endure the side effects—say loss of hair for someone for whom chemotherapy would be highly effective; or of weight gain for someone who needed steroids to treat some condition, for instance. What kind of influence should a health care professional exert?

I was in on a bioethics committee meeting concerning a pregnant woman admitted to the hospital with some sepsis in her uterus. The sepsis was rapidly invasive and life-threatening to both her and her unborn child; it was also easily treatable with antibiotics. The woman, however, believed herself to have a multiple chemical sensitivity syndrome and refused treatment with an antibiotic. The physicians and staff did not accept her claim about her condition and in fact were very skeptical about the existence of any such condition. They were extremely frustrated with their inability to persuade her to take the medication. At one point, she did take an antibiotic but within a few minutes claimed to be experiencing some tingling in her extremities and refused to continue the treatment, a claim discounted by the attending physician who said the antibiotics could not yet have been absorbed into her system. A psychiatric consult was requested and she passed. The physicians obtained a court order to require her to take the medication. They did so on the grounds that the life of the unborn child was at risk and that it was at the point of viability or soon would be. For the most part these were individuals who did not believe that the fetus has full human stature and would allow for the legality of an abortion at any point in the pregnancy. Nonetheless, her situation seemed to them to manifest reckless disregard for the well being of her unborn child. Still, they were hesitant to use the court order and several in fact said they would never have approved of using it but were glad to have it for its "persuasive power." Finally, someone had the bright idea of contacting the patient's homeopathic doctor (the physicians referred to the homeopath as a "guru") who had diagnosed her with this condition. The seriousness of the situation was, of course, pressed upon him and he managed to convince his patient that the antibiotic that she had been prescribed was not one of those chemicals to which she was allergic. She took the antibiotic, the sepsis was treated, and she and baby did well.

Certainly, what her autonomous choice was, was abundantly clear: she did not want to take any medication. But no one was willing simply to honor that autonomous choice. Virtually everyone thought that some methods of persuasion should be used, though some were disinclined to resort to the coercion of the law. It could be argued that it was simply insufficient information that governed her choice and that when she finally heard from her homeopath physician, she was in a position to make a better choice. But the unanimous assessment of those involved was that it was not insufficient information that led to her initial decision nor more adequate information that led to her final decision; rather, her treatment team believed her to be very controlling and desperate for attention and that once she had created enough of a stir and had become the center of attention for a sufficient period of time, she yielded.

One member of the committee fretted that the intervention might have been wrong—and queried whether the team's action had been a violation of the woman's conscience or religious belief, but was eventually satisfied with the argument that erroneous beliefs about one's medical condition are not equivalent to religious beliefs. In the end a large ethics committee with a diverse set of religious and philosophical commitments agreed that the procedure had been proper although most were uncomfortable with the fact that they had violated the woman's autonomy.

Natural law theorists, of which I am one, believe that ultimately everyone reasons by the most basic natural law principles in some fashion; that like the principle of non-contradiction and the rules of logic, they operate in our reasoning whether or not we are conscious of them; certainly we can violate those principles and even attempt to deny them, but we can't truly escape their presence in our reasoning. I think the consensus of this large bioethics committee indicates that no matter what their philosophical or religious commitments and their professed commitment to the principle of autonomy, the human tendency to be a "guiding autonomist" kicked in.

Good friend physicians will make use of a wide range of means of influencing a patient they believe is making a very bad decision. We noted above that Aristotle had a kind of a list for how to teach morality to young people of various temperaments and levels of moral development. Parents and teachers have a wide range of techniques for leading their charges to do what is good. These techniques range from using persuasion, being manipulative, or usually with a sense of last resort, employing coercion or force of some kind.

Faden and Beauchamp provide a very useful and thoughtful description of persuasion, manipulation and coercion. Their descriptions are in themselves value-free; they explicitly note that they believe both coercion and manipulation are, at times, morally justified. In reading their assessments keep in mind that they are concerned to ensure that patients act as autonomously as possible—not that they make a good decision freely (see Faden and Beauchamp, 1986, 339, 354–355). They make a distinction between autonomous action, authentic action, and voluntary or free action; autonomous action is that which is intentional, with understanding, and without controlling influences (238), authentic action is that which is guided by one's reflective preferences, values that are truly one's own (263). They prefer to avoid the terms voluntary and free action since they understand these terms to have too many confusing associations (257), but they do have a category of action that is not autonomous and not coerced and this is the "noncontrolled" action; one that an agent does without coercion but not necessarily with full intention and full understanding. The example given is a patient who takes a drug recommended by a physician largely because the patient believes a better relationship with the physician will ensue, not because the patient is convinced it is the best drug (258).

They state that "Coercion occurs if one party intentionally and successfully influences another by presenting a credible threat of unwanted and avoidable harm so severe that the person is unable to resist acting to avoid it" (339). They include legal sanctions among justifiable instances of coercion; they consider threats to be an instance of usually immoral coercion.

Faden and Beauchamp define persuasion as "restricted to influence by appeal to reason", as "the intentional and successful attempt to induce a person, through appeals to reason, to freely accept—as his or her own—the beliefs, attitudes, values, intentions, or actions advocated by the persuader" (261). They believe that persuasion is morally permissible if not obligatory (347) and always respects the autonomy of the patient (373). They do not provide much justification for their wholesale approval of persuasion; it seems to rest upon an anthropology that understands the human person to be essentially rational and that appeals to reason do not compromise but enhance freedom.[12] (I subscribe to an anthropology that considers our reason to be our highest faculty but which also understands the emotions to be useful receptors of reality and thus proper guides to action on occasion.) The appeal to reason that Faden and Beauchamp advocate is to be done through the use of structured argument and reasoning (they allow for some nonverbal communication such as visual evidence, at 348), and excludes appeals to "emotive reactions and motivations such as hate, fear, disgust, or embarrassment" (351). Yet, I suspect persuasion that uses appeals to reason could seem manipulative or even coercive in some sense to those who do not feel adept at rational debate. Certainly, Faden and Beauchamp define persuasion as getting people to *freely* accept beliefs, etc. but I am not certain that they are fully aware of how intimidating and threatening rational argumentation can be to some people.

Manipulation is defined as "any intentional and successful influence of a person by noncoercively altering the actual choices available to the person or by nonpersuasively altering the other's perception of those choices" (354). Offering incentives of free medications, indicating that emotional support will be withdrawn (354), flattery, and the attempt to induce feelings of guilt (366) are cited as examples of manipulation used in health care. The form of manipulation that concerns them the most is deception, which "uses such intentional strategies as lying, withholding of information, true assertion that omits a vital qualification and misleading exaggeration in order to cause persons to believe what is false" (363). They do not state that they consider all forms of deception wrong, just that deception threatens the kind of understanding that is necessary for informed consent, which, remember, is the goal of their discourse. Another goal, such as getting someone to do what is good, perhaps not as autonomously as Faden and Beauchamp desire, might permit more deception than they are inclined to.

Faden and Beauchamp do not tell us how they evaluate the use of stories to influence others—for instance, the use of the story of someone like Lance Armstrong who through phenomenal effort beat all odds. Are such stories persuasion or manipulation? Do they make an appeal to reason (though not through argumentation) or through emotion (but perhaps an emotion in accord with reason)? Within the last two decades or so, a whole subfield of bioethics, narrative ethics, has emerged. Narrative ethics[13] employs narratives of different kinds (for instance, paradigm cases such as those that feature in legal decisions, comparative cases, inspiring stories and the stories of a religious tradition, and the patient's personal story) to assist the process of decision-making. Narrative ethics can be used to supplement or replace a "rule-based" approach to ethics. Let me suggest that stories that promote something that is possible yet improbable falls into the category of manipulative deception that does not constitute lying or malicious deception but does have an element of falsification to it. The amount of deceit in using such stories is, I suspect, minimal since the patient mostly likely realizes that the model he or she is being motivated to follow is not the norm. The use of such stories, in my view, is not uncommon and, for the most part, moral and wise.

Dealing with patients and moving them to do what is best for them, requires a toolbox of diverse skills, involving various kinds of persuasion, manipulation and coercion. Again, since I subscribe to an anthropology that values reason as man's highest faculty, I believe the preferred method should always be to give the patient all the information that he/she needs to make a good decision and for the physician to explain why he or she recommends a given line of action. If the patient is determined to do something the physician considers to be significantly damaging to his or her health or to be a significant moral wrong, the physician should undertake to persuade the patient to change his/her mind as straightforwardly as possible. But when straightforward rational argumentation does not work, some other alternatives should be considered. The question here is to identify some of those alternatives and to determine when it is moral to use them.

Many physicians and in fact individuals from all walks of life are sometimes willing to attempt to influence someone else who has indicated that he or she is about to make a bad decision (remember the parent attempting to influence the adult child about to marry):

1. When a great harm, physical or psychological, may come to another, innocent party. For instance some individuals may be willing to get involved in the lives of their neighbors when they believe significant abuse is taking place. Physicians may have legal obligations to report abuse.
2. When a violation of the law is involved; some physicians will advise against cocaine not only for medical reasons but also because their patients are risking arrest.
3. When an individual may be risking significant harm to him or her self; for instance, most of us would try to prevent a stranger leaning out a window of a high building and threatening suicide, no matter how rational their reasons for wanting to commit suicide. Others might get involved in situations when much lesser risk is involved especially if we are in some sort of relationship with the agent; we may warn others against bad stock purchases or high caloric food or bad restaurants.

The degree of involvement and the kinds of measures individuals are willing to use is generally governed by various factors such as the type of relationship one has with another; what kind of danger is involved to the person being helped; the kind of risk one is undertaking for one's self or others for whom one is responsible; and the availability and morality of means to be used to exercise the influence one thinks suitable.

The above case of the woman believing herself to have multiple chemical sensitivity syndrome, illustrates some of lines of action one might take; persuasion; coaxing the patient to try something for a little while; calling in an individual the patient respects; having recourse to the coercion of the law. Let me expand upon that list a bit and offer the following.

1. The physician should take special care to make certain that the patient truly understands the consequences of not adopting the recommended treatment. Here I raise a question: would it be wrong to give particular emphasis to the "worse case" scenario if the recommended treatment is not adopted?[14]
   Consider, for instance, the means that we use to persuade young people not to drink and drive and not to smoke. I remember seeing when I was in high school films of people bloodied in car accidents or maudlin tales of young people who died young; I remember seeing pictures of lungs encrusted with carbon and pictures of people with emphysema getting around awkwardly with their oxygen machines. I remember no attempt of those who were warning us not to engage in these practices to make sure we had some accurate sense of what were the percentages or real possibilities of us suffering such dire consequences if we drove under the influence of alcohol or smoked; indeed, efforts were made to ensure that we believed it was almost a certainty that we would suffer such consequences.

Is it not true that when we are attempting to dissuade someone from doing something that would damage them, we make it sound like a 10% chance that something bad will happen to them sound like a very great possibility, and that we are trying to persuade someone to do something that we think would be very good for them, we try to persuade them that a 10% chance that something very bad will happen to them should they undertake the recommended treatment, is a very small chance?

2. Advise the patient to talk to someone who has undergone the procedure. If someone is reluctant to have a limb amputated because of diabetes and is afraid that his/her life will not be worth living, would we not put him/her in contact with someone who has adjusted well to being without a leg rather than someone who has slipped into progressively deeper depression and isolation?
3. Find articles and stories that may inspire the patient to embrace the hardships that may be obstacles to accepting a treatment.
4. Pray and sacrifice for the patient. Perhaps even pray with the patient; ask the patient to pray about the decision. These approaches should be done with delicacy but if the health care professional is a religious person and naturally prays and asks others for prayers, it is fitting that such practices be a part of their interaction with their patients. All the more so, of course, if the patient is religious.
5. Call in the services of a staff member at the hospital who has great bedside manner. These individuals are often able to discover the true fears of patients, to console them, and to cajole them into making the recommended decision.
6. Find a family member, religious minister, or friend who may be able to be persuasive. I have seen physicians frequently do this. If there is a family member who agrees that the treatment recommended by the physician would be best, this family member gets notified first about the patient's condition and gets preferred access to the patient. More obstructionist family members may be excluded gently or forcibly from the consultative process.
7. Request a psychiatric consult. I have seen physicians ask for the consult when it is fairly clear that the patient is certainly functionally sane but stupidly making a bad decision in this instance. Sometimes merely raising the possibility of ordering a consult shames the patient into succumbing.
8. Get a court order if feasible. Here I am not suggesting that the law is the final arbiter of morality but on occasion it can help us pursue moral choice making.
9. For the sake of the completion let me raise the question of lying to a patient. Let me state that I think saying something false so as to deceive the patient about a serious matter would only be right in the most extreme circumstances.

Here I cannot go fully into a justification, but let me say that I think parents, spouses, teachers, physicians, fund-raisers, and politicians and perhaps health care professionals say false things with some regularity and perhaps some justice. I knew a child who was willing to eat only what he thought was chicken and thus his parents labeled everything he was served as "chicken." Many a spouse has said "no you don't look fat, dear" or "No, I don't mind your buying those season tickets" when the truth was otherwise. Teachers say "I know you can do it," when they have great doubts about a student's ability to perform. Fund-raisers guarantee they will make their goals, and politicians are politicians. Physicians routinely say "this won't hurt a bit, or much..." And then we have Jews equipped with a false passport by the Vatican as they attempt to escape the Nazis, or Father Cizek who lied about his marital status when doing underground priestly work in communist Russia.

I do not think that to justify the saying of what is not true in some circumstances is to commit necessarily to "consequentialism," or to allowing the circumstances to completely define the morality of a choice, to the extent of being willing to do something evil[15] to achieve what is good. The quick but insufficient way of explaining why I think the above communications are moral is that I believe in all of these relationships there is some kind of "code" being spoken: spouses sometimes don't want or expect the truth from their spouses; students sometimes need encouragement not truth from teachers, etc.; hostile border guards in time of warfare or oppression do not expect the truth either. Extreme circumstances in the medical profession might be telling someone who has been severely injured in a car accident and inquires about her young son who was also in the accident, that "he is doing fine, don't you worry" when perhaps he is not doing fine and a rational person would definitely worry.[16] Here I think a kind of "code" is understood. Extreme circumstances would not include lying about a fetus's gestational age to prevent a woman from getting an abortion. There is no code that is governing this situation; rather it requires the utmost commitment to truth.

It is my impression that health care professionals regularly employ such techniques as listed above (and undoubtedly more that I have not identified), even those who profess that they have a profound respect for autonomy and those too who think there are no objective moral norms.

Let us return to the case of the Jehovah Witness who refuses a blood transfusion for him/herself or for a child.[17] At one time our culture had significant moral consensus and was willing to override the autonomous choice of the Jehovah Witness at least in respect to the treatment of their children; now we are tending to extend the protection accorded to matters of conscientious "religious belief" to the refusal of medical treatment of any kind (and even the request for procedures such as transforming someone into a lizard or amputating a healthy limb).

Theologians in the Catholic Church who understand the Church to hold to the set of philosophical commitments I outlined above, universally agree that it would be wrong to force an adult Jehovah Witness (one not pregnant) to have a blood transfusion and it seems for two reasons; one is an appeal to the universal right to refuse any and all medical treatment and the second is from respect for the deeply held religious views of others: we should not prevent people from acting on their religious beliefs. (Though we do prohibit some religious practices such as smoking peyote.)

I am not convinced that either is a sufficient justification for not trying as hard as one can to convince the Jehovah Witness to get a blood transfusion. And this is not because I value life over religious belief. Although I truly believe that Jehovah

Witnesses who die because of a refused blood transfusion are not endangering their souls, and in fact, may be ensuring their salvation because of their willingness to die rather than violate God's will (and I believe the same of those who crashed into the World Trade Towers, if such was their motivation), I do not believe that is a reason necessarily to leave their choices unchallenged.

Here, again, attention to the particular is of great importance; if the physician perceives that a patient desperately wants to live and would greatly appreciate the physician overriding his/her express wishes, perhaps the physician should in fact do so, if there are legal remedies for achieving such (as there was for the woman with the sepsis). This would especially be the case if one had the support of the family, one's colleagues, one's institution, and the law. On the other hand, another situation might call for a very different solution. Say there was great enmity between one's own religious group and Jehovah Witnesses. Suppose doing the procedure would greatly exacerbate the relationships between the groups and lead to the Jehovah Witnesses becoming very hostile to the truth that one wishes to promote. Suppose one's colleagues would disdain one and one's institution would fire one or that one would be prosecuted. In this instance, not pushing for the transfusion may well be the moral thing to do.

I once heard a Catholic theologian, very highly placed in the Vatican, maintain that as long as a Jehovah Witness was conscious one had the obligation to respect his/her refusal, but if he/she slipped into unconsciousness, one could perform the procedure, under the assumption that one is now the caretaker for the patient and must do what is the true good for the patient. This might be equivalent to refusing to honor the wishes of someone who one knew intended to bequeath his three million dollar estate to build a mausoleum for his cat; should he die before those provisions are set in a will, and should you be the executor of the estate, you would be morally justified, if not obliged, to ignore those expressed wishes. Now I do not mean this analogy to suggest an equivalence between the wishes of the Jehovah Witness and the wishes of the cat-owner; I mean rather to show that once one becomes the decision maker, the principle of "substituted judgment," the principle that states that we must decide how the individual would have decided, may not always be the principle to be honored.

In spite of the line of reasoning set forth above, I am very uneasy about advocating that health care professionals override the decisions of others in respect to their personal decisions and certainly health care qualifies as one of those decisions. I am uneasy for several reasons:

1. Human beings are by nature self-determining; to interfere with that self-determination threatens to interfere with what is most deeply their own.
2. The good that is to be obtained by good moral decisions, i.e., a good moral character, is only to be obtained through free decisions, not through coerced decisions and, in my view, the acquisition and retention of a good moral character is the foremost good.
3. There are many subjective factors that may rightly influence an agent's decision and it is sometimes very difficult or inappropriate for the agent to articulate

those reasons, so it is sometimes impossible for another to know the real reasons governing an agent's decisions. Women, for instance, differ radically in their choices for treatment of breast tumors; one woman with precancerous tissue may decide to go for a radical or double radical mastectomy whereas another may choose to forego treatment until cancer develops.
4. No individual is infallible in moral decision-making and thus should rightly be hesitant to risk imposing a wrong moral judgment on another.
5. We live in a culture in which moral views are widely diverse; for instance, a Catholic would not want a secular humanist making medical decisions for him/her and vice versa.

Thus, there are several reasons why, even if there is not an absolute right to refuse medical treatment and even if autonomous decision-making is not the highest good, great care must be taken with any decision to exercise strong influence over a patient's decision and even more so when a patient has made a carefully considered autonomous choice.

This essay has obviously been written from the perspective of the health care professional. Clearly, the perspective of a patient about these matters may be altogether different and would be the subject of another essay. We rightly value our autonomy, we wish to be self-determining and to manage our lives in accord with our own values. Yet we also welcome to some extent the protectiveness of government that involves regulating medical treatments and procedures. This paper alleges that the bioethics has not fully reconciled its advocacy of the principle of autonomy with standard medical practice and needs to engage more self-consciously in the consideration of kinds of influence they may morally employ and when it is moral to employ them.[18]

Notes

1. I am situating my position somewhere between Tristram Englehardt's concept of "moral strangers," developed in his Engelhardt (1996), and that in Wildes (2000) (my unpublished review of Wildes' book is available on my website: http://www.aodonline.org/SHMS/Faculty+5819/Janet+Smith+9260/Dr.+Janet +Smith+-+Book+Reviews.htm).
2. I will be using "physician" interchangeably with "health care professional": I mean my remarks to apply more widely than to just physicians; I use "physicians" for the sake of economy.
3. Others have made this point well; see, for instance, Ackerman (1982).
4. I have written about this elsewhere; see Smith (1997).
5. It is not unusual for bioethicists to invoke the principle that health care professionals have on occasion an obligation to influence the decisions of their patients, but do not often justify that principle.
6. See for example, Ahronheim et al. (2000); Crigger (1998); Freeman and McDonnell (2001).

7. I had nearly completed this essay when I rediscovered the excellent definitions of these terms and descriptions of various instances of them in Chapter10 of Faden and Beauchamp (1986).
8. A good review of the legal right can be found in Malloy (1998).
9. For instance, Germain Grisez states, "Except in certain special cases provided for by law (for example, immunizations to prevent epidemics), no sort of health care ever should be imposed upon competent adults without their actual or reasonable presumed consent" (Grisez, 1997, 330). Joseph Boyle has defended this claim in Boyle (2002).
10. When Aristotle speaks of finding the "mean" or midpoint to determine what is right, he gives the example of the athletic trainer, who knows that the "mean" of the amount of meat needed for Milo would exceed that average mean, since Milo is larger than others (Apostle, 1984, 1106b).
11. See http://www.thelizardman.com/.
12. Some of the fundamental philosophical commitments that drive Faden and Beauchamp's analysis are advanced without philosophical justification and explicitly so. For instance, in defending the principle of autonomy they do not provide anthropological justifications for it but state: "...the burden of moral justification rests on those who would restrict or prevent a person's exercise of autonomy" (Faden and Beauchamp, 1986, 8).
13. For a good introduction to the use of narrative in bioethics see Nelson (1977).
14. Faden and Beauchamp call these "fear appeals" (Faden, and Beauchamp, 1986, 351).
15. Here is not the place to rehearse the debate within Catholic moral theology between intrinsic evil and premoral or ontic evil. I have written about this matter elsewhere. See Smith (1998, 1999).
16. Certainly some will say that these statements by a physician are not lies; that one could be meaning "fine for someone who was in such a terrible accident" and "you shouldn't worry because you need your strength to recover" but the words are being used against their accepted sense, with the intention to deceive, so I think lying is the right description for the action.
17. The classic article on this issue is Macklin (1977); a more updated review of the issue can be found in Ahronheim et al. (2000, chapter 7).
18. A version of this paper was delivered at the annual bioethics conference at Wheaton College, Oct. 31, 2003; I thank Francis J. Beckwith for his useful comments.

References

Ackerman, T.F. 1982. Why doctors should intervene, *Hasting Center Report* 12:14–17.

Ahronheim, J.C., J.D. Moreno and C. Zukerman. 2000.*Ethics in clinical practice, 2nd edition*. Gaithersburg, MD: Aspen.

Apostle, G.H., Trans. 1984. *Aristotle's nicomachean ethics*. Grinnell, IA: Peripatetic.

Boyle, J. 2002. Limiting access to health care: A traditional roman catholic analysis, In *Allocating scarce medical resources: Roman Catholic perspectives* eds. H.T. Engelhardt and M. Cherry, 77–95. Washington, DC: Georgetown University Press.

Crigger, B. 1998. *Cases in bioethics, 3rd edition.* New York, NY: St. Martin's Press.
Elliott, C. 2000. A new way to be mad, *Atlantic Monthly*, Dec. 2000. Available on-line at: http://www.theatlantic.com/issues/2000/12/elliott.htm. Accessed 04 Sep 2010.
Engelhardt, H.T., Jr. 1996. *The foundations of bioethics, 2nd edition.* New York, NY: Oxford University Press.
Faden, R. and T. Beauchamp 1986. *A history and theory of informed consent.* New York, NY: Oxford University Press.
Freeman, J.M. and K. McDonnell 2001. *Tough decisions: Cases in medical ethics, 2nd edition.* New York, NY: Oxford University Press.
Grisez, G. 1997. *The way of the lord Jesus: Difficult moral questions* vol. 3. Quincy, IL: Franciscan.
Macklin, R. 1977. Consent, coercion, and conflict of rights. *Perspectives in biology and medicine* 20:360–371.
Malloy, S.E.W. 1998. Beyond misguided paternalism: Resuscitating the right to refuse medical treatment. *Wake Forest L. Review* 33:1035–1093.
Nelson, H.L. 1977. *Stories and their limits: narrative approaches to bioethics.* New York, NY: Routledge.
Ross, W.D., Trans. 1979.*Aristotle's Metaphysics.* Oxford: Oxford University Press.
Smith, J. 1997. The pre-eminence of autonomy in bioethics, In *Human lives: Critical essays on consequentialist bioethics* eds. D. Oderberg and J.A. Laing. London: McMillan 182–195.
Smith, J. 1998. *Veritatis splendor*, proportionalism, and contraception. *Irish Theological Quarterly* 63:307–326.
Smith, J. 1999. Moral terminology and proportionalism, In*Recovering nature : Essays in natural philosophy, ethics, and metaphysics in honor of Ralph McInerny*, eds. T. Hibbs and J. O'Callaghan, 127–146 Notre Dame, IN: University ofNotre Dame Press
Wildes, K. 2000. *Moral acquaintances: Methodology in bioethics.* Notre Dame, IN: University of Notre Dame Press.

Part V
The Right to Health Care

Chapter 10
Social Justice, Charity and Tax Evasion: A Critical Inquiry

Mark J. Cherry

10.1 Introduction—The Welfare Entitlement Claim

Through many conversations, in several countries, multiple states, and on at least two continents, Joseph Boyle has with great generosity and philosophical clarity taught me natural law theory. It is an honor to contribute to this volume. More importantly, the opportunity to grapple with his arguments is always rewarding. His defense of the new natural law theory is inevitably forceful and often almost convincing. As Boyle has noted, natural law theory provides a sophisticated interpretation of what Alan Donagan termed "common morality": that part of the Judeo-Christian moral tradition that is non-religious. It is comprised of moral precepts guiding action that are secular, rather than religious, in character: "its subject matter is actions affecting oneself and other human beings, not God" (Boyle, 1999, 111). These precepts are claimed to be knowable through general human reason without particular religious revelation or theistic conviction (Boyle, 1999, 111; Donagan, 1977). Natural law theory espouses universal moral principles and precepts or norms, but not in a way that invokes God, religion, or immortality, to direct and motivate moral action. As Boyle encapsulates the position:

> I understand "natural law" to be the name of the set of basic principles of moral life. These principles are prescriptive and so constitute a "law." They are accessible to the intelligent reflection of any human being capable of thought and action and in that sense "natural" and "universal." Natural law theory is the elaboration of these principles into a normative account of moral thought and life. Roman Catholic canonists, theologians and philosophers have developed natural law theory within a tradition of moral inquiry since at least the 12th century. The result of this trans-generational inquiry is a developed form of traditional morality (2006, 300).

Natural law holds that objective moral truth is not the product of cultural practice, social convention, individual choice, or even religious conviction. Rather, natural law's moral content is held to be universally applicable—its moral content is validly

M.J. Cherry (✉)
St. Edward's University, Austin, TX, USA
e-mail: markc@stedwards.edu

C. Tollefsen (ed.), *Bioethics with Liberty and Justice*, Philosophy and Medicine 110, DOI 10.1007/978-90-481-9791-0_10, © Springer Science+Business Media B.V. 2011

addressed to anyone facing a choice upon which the principles and norms bear—and universally accessible—that canonical moral content is accessible to human beings generally (Boyle, 2006, 2). All persons are in principle capable of knowing canonical moral principles and norms of action in virtue of their common capacity for practical reason. Natural law's moral epistemology assumes that persons are capable of appreciating, and should be held responsible for, the universal moral truths revealed through the underlying basic goods of human practical reason.

Boyle describes the categories of basic human goods as including health and safety, knowledge and appreciation of beauty, excellence in work and play, as well as friendship, peace and fraternity.[1] Among his conclusions regarding the universal moral truths of the natural law, which he argues follow from such basic goods through human practical reason, are significant welfare entitlement rights as well as for corresponding governmental duties to provide for such entitlements—that is, for considerable state-based redistributive taxation to pay for wide-ranging welfare entitlements. He argues that "political society should use its legal authority to guarantee that the welfare rights of its needy members are recognized" (2001, 207). Welfare entitlement rights

> are not simply the claims of some and the duties of others to forbearance, but claims and duties to provide or guarantee the provision of a fair share of those instrumentalities or empowerments important for living good human lives: these include things like education, housing, health care, and welfare support for the elderly, the unemployed, and the underemployed (Boyle, 2001, 206).

He holds that citizens ought to understand such taxation as morally permissible and indeed, as part of their general moral obligation to assist the needy. "By requiring owners to pay taxes, political society enforces this right of the needy against owners who might otherwise refuse to act responsibly; in this way, political society vindicates the rights of the needy and gives their entitlements legal form" (Boyle, 2001, 224). He thereby seeks to place such extensive welfare entitlements within the boundaries of legitimate moral political authority.

This paper critically explores Boyle's underlying moral and social political claim: that the welfare needs of persons generate politically enforceable welfare entitlement rights. As I will argue, that some person needs a good—even to avoid suffering and premature death—does not thereby give that person a moral entitlement to the good in question; nor does such a need necessarily establish a corresponding moral duty that others satisfy the need. Moreover, that an action, i.e., redistributive taxation, promotes a special good, such as health or human flourishing, or fulfills a welfare need does not thereby make the action morally permissible.

10.2 Initial Considerations

Consider some actions that promote special goods or that fulfill welfare needs, but which are widely judged illicit—even by natural law theory. Moral norms regarding assault and battery (unauthorized touching or use), for example, prohibit professors from forcibly immunizing their students against influenza, even though this would promote the special goods of health and life, both for the students themselves and for others, such as the professor, with whom the students regularly come into contact.[2] Similar restrictions regarding the nonconsensual use of others likewise prohibit professors from surgically removing each student's extra kidney even if the professor will arrange for the organs to be provided to patients, who need a transplant and will suffer and die without the necessary organ (see generally Cherry, 2005). Similarly, norms regarding theft generally prohibit professors from redistributing their wealthier students' new books to their students in need, even when the wealthy student routinely skips class, while the student in need is diligent. Persons may fail fully to utilize their property, or to utilize their property in ways of which others approve, but neither circumstance establishes the moral claim that such property ought to be redistributed to others, or that those in authority—in this case, professors—have the legitimate moral authority to redistribute.

Just as there are moral limits on legitimate personal or professorial behavior to promote special goods, there are moral limits on legitimate governmental regulatory action to promote special goods, such as health, housing, education or other welfare entitlements. A core question of any governmental act concerns whether it is a legitimate application of moral political authority or an unauthorized act of state bureaucratic coercion, regardless of whether the state action creates or promotes a special good, such as health care, or fulfills the welfare needs of persons.[3] For example, state laws mandating the use of seat belts in automobiles may be determined to be illegitimate and immoral acts of coercive force, even if such laws successfully save lives.

10.3 A Natural Law Argument for the Legitimacy of the Welfare State

10.3.1 Step 1: Individual Duties to Others and the Creation of Welfare Rights

Boyle's argument begins with the background moral assumption that persons are obliged to help those in need. His appeal to authority relies on Thomas Aquinas: "Catholic exponents of natural law, at least since Thomas Aquinas (A.D. 1224–1274) have emphasized the common obligation of every person, as his or her capacity allows, to assist those in need" (Boyle, 2001, 207). Boyle acknowledges that charity is an imperfect duty: it is limited both by individual capacity as well as by being a responsibility "...at least initially, of individuals, and perhaps of small communities like families, neighborhoods, or parishes—it is not a responsibility of political society" (Boyle, 2001, 207). Aquinas treats giving alms to the poor as a moral obligation (Aquinas, *ST* II-II, Q.32, a.5). Boyle similarly concludes that the natural law requires property owners to use their "residual or superfluous" to assist those in need: "the natural law requires owners to use anything residual or superfluous to help the needy" (Boyle, 2001, 212). Leaving aside, for the moment, the challenge of producing a non question-begging definition of "residual or superfluous," Boyle has only concluded that there is a moral obligation to give to those in need from our "residual or superfluous" as an act of private charity. Thus far, the argument has established neither that the poor generally have welfare rights, nor that any particular person has a welfare right, to receive any particular set of goods or services, from any other specific person or group of persons.

The argument to establish welfare rights is as follows: "...if those capable of helping a needy person have a duty to help that is grounded in the welfare of the needy person, then the needy person has a right to that help grounded in need" (Boyle, 2001, 215). To tease out the specifics of the argument, consider a restatement:

(1) A is capable of helping B, who is in need, from A's residual or superfluous.
(2) If A is capable of helping B, who is in need, from A's residual or superfluous, then A has a duty to help B, grounded in B's need.
(3) Therefore, B has a positive entitlement right to welfare assistance from A, grounded in B's need.

Assuming that premises 1 and 2 are plausible, the philosophical work comes in the move to the conclusion. The challenge, though, is that the conclusion announces a welfare entitlement right, whereas the premises stipulate only a limited imperfect duty. To establish the conclusion, Boyle needs to demonstrate a necessary connection between A's *duty* to assist those in need and the affirmed *right* of those in need to receive the assistance. Perhaps A has a duty to give to someone or other, who is in need, but no particular person in need has a right to receive A's charity.

Boyle provides part of the necessary connection in a footnote—Joseph Raz's definition of a right: "*Definition*: 'X has a right if and only if X can have rights, and, other things being equal, an aspect of X's well-being (his interest) is sufficient reason for holding some other person(s) to be under a duty" (Boyle, 2001, 215, fn). This key premise unlocks a correlation between a duty to assist those in need and the entitlement right of those in need to receive the assistance. It is worth noting, though, that even if we grant this definition of a right, it has not yet established that any particular person in need has a right to be given charity from any other specific person's residual or superfluous; i.e., B might have a right to charity, but A might not have a duty to provide such charity to B. In such circumstances, it would be morally illegitimate to coerce A to provide charity to B.

10.3.2 Step 2: The Move from an Individual Obligation to a Social Obligation

Boyle next moves from an individual duty of charity to voluntary coordinated social action to help the needy. The moral argument is grounded in an appeal to efficiency and effectiveness:

> The social organization of voluntary groups to help the needy appears to be common in most societies. Moreover, on natural law grounds, an individual's cooperation with such organizations by joining them and giving them resources is frequently more than a voluntary and charitable action—it is an obligation. *Such an obligation exists when a person is obliged to provide assistance, but cannot do it (or do it well) without cooperating with others. Thus, although the underlying obligation is, and remains, an obligation of each owner, common action is often the most efficacious course in modern society* (Boyle, 2001, 220, emphasis added).

The argument is relatively straightforward:

(1) Each individual has an obligation to give to the needy.
(2) Cooperating with others through social organizations, such as voluntary groups, is the most efficacious way of providing for the needy.
(3) We ought to help others in the most efficacious manner.
(4) Therefore, there is an obligation to provide charity to the needy through voluntary social organization.

Here, the argument moves beyond individual imperfect obligations of charity to help others out of one's residual or superfluous, to a moral obligation to help those in need *in the most efficacious manner*—that is, to "do it well." So, even if I can assist others reasonably well through personal private charity, if such assistance can be accomplished in a more efficacious fashion through some form of social coordination, then my purported moral obligation is to work towards such social coordination. Whereas it may be true that we can more efficiently and effectively assist others through voluntary social groups, that we ought to cooperate with others most efficaciously to provide for the needy, adds an additional moral assumption. The plausibility of this assumption in turn rests on the reasons grounding the obligation to be charitable. What are the core goals of charity? A foundational moral account of charity must be given.

10.3.3 Step 3: The Move from a Social Obligation to Coercive Political Authority

The final step in the creation of the welfare state places the first two conclusions regarding welfare rights and efficacious social organization within an argument regarding legitimate use of the coercive power of the state. This move is necessary if natural law theory is to secure a general welfare entitlement from the imperfect duty of each person to give to charity.

> The social responsibility of owners, I have noted, is rooted in their having a surplus and in the welfare of the needy; even before there are any voluntary or political initiatives to assist them, the needy have a moral right to the help that can be given to them.. . .Thus, political society does no wrong to irresponsible owners who resent or seek to evade the tax, since their pre-political duty to the needy already included giving what reasonable taxation requires (and probably then some). Nor does it wrong the law-abiding, who accept this political action as a needed way of coordinating the administration of an important responsibility connected to their membership in political society. Indeed, the need to coerce the

> irresponsible, in order to vindicate the rights of those harmed by their irresponsibility and to secure for the law-abiding the benefits of their own social responsibility, is an important reason why the development of welfare rights is a proper political action (Boyle, 2001, 224).

Coercive political authority makes it possible to combine the enforcement of (1) the moral obligation to give to the needy with (2) the efficaciousness of social distribution that is possible with large scale social coordination. State-based coercion enforces the welfare entitlement rights of the needy: "By requiring owners to pay taxes, political society enforces this right of the needy against owners who might otherwise refuse to act responsibly; in this way, political society vindicates the rights of the needy and gives their entitlements legal form" (Boyle, 2001, 224). Here the government both recognizes and enforces welfare rights as entitlement rights. In Boyle's terms, the government is simply forcing persons to fulfill their moral obligations, which they ought to have done anyway.

10.4 Analysis

Among its central assumptions, this natural law argument for state-based welfare entitlements stands or falls on (1) the definition of a right, (2) the claim that welfare needs create politically enforceable welfare entitlement rights, and (3) that efficaciously meeting the welfare needs of others is, or ought to be, the moral obligation of charity. If any, or all, of these three core assumptions is found unreliable or defeasible there will be sufficient grounds for concluding that the argument does not ground welfare entitlement rights, much less coercively enforceable welfare entitlement rights. Such a conclusion will be sufficient to block the move to the welfare state because governments will not possess the legitimate political authority to engage in coercive redistributive taxation to enforce either any claimed pre-political imperfect duty of charity or any supposed requirement that charity distribution in fact be efficacious.

10.4.1 Assumption 1: What Does it Mean to Have a Right?

Rights language is deeply morally and philosophically ambiguous. That some person A has a right to B might be understood in terms of any number of possible definitions:

(1) All things considered A has a right to B: i.e., *ceteris paribus* A ought to be left alone in the enjoyment of B—the right is a *prima facie* claim, perhaps strong, but which can be overridden by countervailing considerations[4];
(2) Rights as side-constraints—regardless of the consequences, A ought to be left alone in the enjoyment of B—the right functions as a side-constraint on the actions of others, including governments[5];
(3) Rights as defining a sphere of moral jurisdiction—A has a right to do as he wishes vis-à-vis B, perhaps, but not necessarily, constrained by other considerations[6];

(4) Goal directed rights—A having a right to B maximizes, or otherwise promotes, the favored end-state or good, or leads to more benefits than harms, however benefits and harms are defined—John Stuart Mill, for example, attempted to place moral rights within a utilitarian moral framework, others have aimed at wealth maximization[7];
(5) Rights as positive entitlements: A has a right to B means that some other person or government has a corresponding duty to provide A with B;
(6) Rights as correlatives of obligations, whether or not the obligation involves a positive entitlement: "X has a right to Y against Z, just in case Z has an obligation to X not to deprive X of Y or to withhold Y from Z" (Brody and Engelhardt, 1987, 11).
(7) Rights as indicating a good; i.e., it would be good if A obtained B—here the language of a right is largely rhetorical; it is a linguistically forceful way to set out a moral or political agenda[8];
(8) Legal rights: A has a legally enforceable entitlement to B, whether or not A has a moral entitlement to B—legal rights might be created either through personal agreement, such as a contract, or legislative enactment.

In addition, rights have been conceived of as (1) alienable, (2) inalienable, (3) created by human action or (4) discovered in human nature; there are (5) claim rights, (6) forbearance rights, and (7) quasi-forbearance rights. Neither of these lists is intended to be all encompassing; other conceptions of rights are possible.

Rights usually correlate with a duty, but not necessarily to a duty to provide positive welfare entitlements. Consider the frequently cited "right to life." If persons have a "right to life" as an alienable side-constraint, then while no one is obliged to keep you alive, others have a duty not to kill you absent your permission. So understood, a "right to life" would not constitute a duty that others provide you with healthcare, housing, food, or emergency rescue. More detailed analysis is necessary to assess under what circumstances individuals could be licitly killed without their permission, such as killing in defense of self or others, killing in just war, or capital punishment in obedience to the law of ben Noah to spill the blood of a murderer.[9] In contrast, if one has a "right to life" as an inalienable positive entitlement, then others might have a duty to keep you alive, perhaps even at considerable expense. Legal "rights to life" vary by content according to jurisdiction and goal directed "rights to life" vary in meaning according to the preferred end-state. For example, if the goal is utility maximization then the meaning of a "right to life" depends on the costs and benefits of keeping particular persons alive, given ways in which resources might otherwise be utilized so as best to create utility. Other types of rights can be created as hybrids, by choosing among the various possible characteristics.

As noted, Boyle's argument for positive welfare entitlement rights relies on Raz's definition of a right: "*Definition*: 'X has a right if and only if X can have rights, and, other things being equal, an aspect of X's well-being (his interest) is sufficient reason for holding some other person(s) to be under a duty." Given the many other possibilities for understanding a "right" it is unclear why we should accept this particular definition, especially since it appears to smuggle in robust content-full moral content by defining a connection between the *duty* to assist those in need and the affirmed *right* of those in need to receive the welfare assistance.

10.4.2 Assumption 2: Do Welfare Needs Create Welfare Entitlement Rights?

Absent prior agreements or particular content-full moral commitments, needs do not generate enforceable welfare entitlement rights. Consider Hillary, who is suffering from kidney failure and needs a transplant. Healthy, well functioning organs are natural advantages, redistribution of which would help to equalize among persons so that each would be able more equally to share in the benefits of society. The 5-year survival rate for those who receive kidney transplantation is excellent, while life expectancy for donors is not measurably decreased. Redistributing a kidney from Bill to Hillary would be at most a temporary inconvenience for Bill, which would greatly enhance Hillary's life expectancy, together with her personal and political prospects. Yet, by itself neither Hillary's need nor the likely positive outcomes establishes Hillary's right to a transplant. Nor do such circumstances establish that Bill does something morally wrong when he fails to donate his redundant kidney to Hillary. Such a circumstance holds even though Hillary may suffer on dialysis and die sooner than she would with a transplant from Bill. We might judge persons who choose to donate a kidney to have acted charitably or to have acted in an exemplary fashion, but neither judgment entails that those who fail to donate act wrongly. Even though there is a very real sense in which Bill's second kidney is redundant (i.e., it can be understood as superfluous)—this is what makes living kidney donation possible—to establish the very strong claim that those who fail to donate act morally wrongly, much more must be said.

Here, prior agreements and background moral circumstances might matter morally: perhaps Bill has a special duty to donate a kidney because Hillary is his wife; but if they are estranged from each other, such a duty might be less plausible. Or, perhaps, Bill has signed a contract with Hillary, agreeing to sell his kidney to her, creating a special legal contractual duty.[10] If they are strangers, however, and Bill hears about Hillary's need for a kidney transplant on the evening news, it is not at all clear why he would have a moral duty, much less a politically enforceable obligation, to donate a kidney. What reasons could one give to establish that Bill is obliged, or ought to be politically forced, to submit to a potentially dangerous surgical mutilation of his body? Again, much more may be said, but this much is clear: neither Hillary's need for a transplant nor the fact that she will suffer and die without a transplant is sufficient in itself to establish either the duty to donate or the moral political authority legitimately to coerce Bill to submit to the surgery. Hillary's need for a transplant does not create an enforceable welfare entitlement right to receive a transplant.

Even if natural law theory could establish more generally that at least some persons have welfare rights grounded in need, these may still be welfare rights that no one in particular has a duty to satisfy. Consider, Bill, Albert, and George, each of whom has a duty to give out of their residual or superfluous to those in need. Each gives 100% of their extra to Richard, who is in need. Hillary arrives late, but is also in need. On Boyle's account Hillary has a welfare entitlement right. However, since Bill, Albert and George have each already fulfilled their individual duties of charity, none has any further moral obligation to give to charity. Absent prior agreements or specific person-relative duties, such as duties to one's spouse or child, Hillary's welfare right does not hold against any particular person; thus, no one is obliged to provide Hillary with the necessary goods or services to fulfill her need.

Such circumstances are certainly among the motivations for moving from a system of private charity to the presumably more efficacious system of voluntary organizations. A significant challenge, however, is that different voluntary organizations will organize private charity to achieve perhaps widely divergent accounts of the good, given competing background moral interests. For example, the Roman Catholic Church might organize charitable healthcare that prohibits abortion. Planned Parenthood might organize charitable healthcare to encourage unfettered access to abortion services. There are Moslem charities and Jewish charities, Orthodox Christian charities, and a wide variety of fully secular charities—each with its own background moral assumptions and social aspirations. The underlying moral character of voluntary charitable organizations will likely mirror the moral pluralism of the underlying society. Individuals will need carefully to choose to whom to give charity and from whom to receive charity. Such choices are not neutral. Boyle, for example, is unlikely to appreciate giving charity to Planned Parenthood in support of abortion for the impecunious to be a virtuous way of fulfilling one's pre-political charitable duty. Similarly, supporters of Planned Parenthood are likely to appreciate Boyle's donations to his local Roman Catholic parish as being subtly subversive of the reproductive rights of women.

The move from voluntary charitable organizations to governmental taxation and welfare distribution is even more problematic. There are no satisfactory grounds for believing that governments are good judges of what constitutes a citizen's residual or superfluous, which each owes to charity. Moreover, those citizens who have already fully satisfied their personal charity obligations, will be morally wronged by the additional burden of governmental taxation, even to satisfy the welfare needs of others. In addition, when governments establish welfare entitlements they instantiate and enforce a particular view of the good, a particular moral content. Which needs quality as those that the state ought morally to satisfy through its redistributive programs?

To illustrate, again consider the heuristic of abortion. Should public welfare policy begin from the anti-abortion position, or from a pro-abortion position? Should public policy begin from the position of Traditional Christian truths or from an atheistic secular cosmopolitan worldview, which insists that no such deep truth exists? Neither is neutral. In *Planned Parenthood v. Casey* (505 US 833, 857–859 [1992]), the court argued that in *Roe v. Wade* (410 US 113 [1973]) the significance of individual control over the self and to personal bodily integrity had been correctly applied to a woman's right to choose abortion. Concerns to preserve the unborn living fetus, even in cases of late-term partial birth abortion, failed to satisfy the burden of proof, according to the court, to override the significance of individual authority over one's body. Respect for individual freedom, the court held, is of greater significance than human life. Indeed, nearly unfettered access to abortion, often through state-based taxpayer financing, has been endorsed as central to preserving human rights, as liberating women from the alleged evils of patriarchy and enforced pregnancy. For example, the United Nations guideline, "HIV/AIDS and Human Rights" (1996), specifically calls for easy access to abortion: "Laws should also be enacted to ensure women's reproductive and secular rights, including the right of . . .means of contraception, including safe and legal abortion and the freedom to choose among these, the right to determine the number and spacing of children. . ."[11] Failing to provide safe and legal abortion as a basic welfare entitlement right is perceived as a violation of basic human rights.

It is only too evident that political struggles concern not merely which policies will best achieve desired objectives, but which objectives are themselves desirable—that is, which moral understanding should be established (e.g., pro-life or pro-choice). How then should public policy be crafted? Should one simply acquiesce to the personal preferences or deep moral intuitions of academic ethicists, current moral convention, or claims to global moral consensus? Or should one seek moral content to guide public policy through appeal to intuitions, consequences, casuistry, the notion of unbiased choice, game theory, middle level principles, or the natural law? All such attempts, as H. Tristram Engelhardt, Jr. argues, confront insurmountable obstacles: one must already presuppose a particular morality so as to choose among intuitions, rank consequences, evaluate exemplary cases, or mediate among various principles; otherwise one will be unable to make any rational choice at all. As he points out, even if one merely ranks cardinal moral concerns, such as liberty, equality, justice, and security differently, one affirms different moral visions, divergent understandings of the good life, and varying senses of what it is to act appropriately.

Such circumstances are not overcome through an appeal to the "so-called" basic goods of the new natural law, or appeal to practical reason, because one must assume a robust moral account regarding which goods are basic, the nature of their content, and how they are to serve as the basis of practical reasoning. One must also assume an account of how to reason from the basic goods to more robust moral conclusions, such as the moral political authority to establish welfare entitlements. Even appealing to general moral frameworks for decision-making, such as the Golden Rule, will fail because understanding the Golden Rule requires specification of particular moral content. Consider: some argue that social beneficence commands all to participate in the establishment of a totally encompassing welfare state, including universal single-tier governmentally regulated healthcare, with tax-payer financed abortions for the impecunious (such as Canada),[12] or tax-payer financed gender reassignment surgery (such as Germany) (Schmidt, 2002). Those who disagree are to be coerced into participation. Others argue that beneficence requires understanding persons as free to refuse to be complicit in what each sees as evil, with basic welfare needs, such as health care, limited to peaceable consensual interaction, including contractual relationships and acts of private charity. Why would "charity" necessarily command that all accept and participate in embryo experimentation and human cloning, or that all contribute through taxation to its development? The challenge is such that because of divergent understandings of the good life and of what actually counts as doing good, general secular reason cannot comprehend beneficence as the Golden Rule.

Tax-payer financed, state-based welfare rights, such as health care, routinely require individuals to participate in the provision of services that they find to be morally objectionable. In such a system, rights not to participate in the propagation of services that one conscientiously opposes, to utilize private property only in ways that one believes to be morally acceptable, and to give to charities with whom one shares moral goals and aspirations, are, at best, insecure and limited. Such losses are not negligible. Traditional Christians are coerced into supporting that which they know to be gravely evil: offering, providing, and financing abortion, physician-assisted suicide, in vitro fertilization, sex reassignment surgery, and so forth. Private property and private charity enhance one's liberty to create one's own understanding of the good: they minimize the force that others can rightfully use against you and require that persons deal with each other through mutual respect and deliberation.[13] In contrast, state based taxation simply takes by force; moreover, taxation to support what one understands to be deviant moral visions significantly impoverishes one's own liberty. Taxation is a powerful weapon for suppressing moral disagreement and for privileging a particular moral viewpoint. As Chief Justice Marshall noted in *McCulloch v. Maryland* (1819), the power to tax is the power to destroy.[14]

So, whose account of "do unto others what you would have them do to you" (Matthew 7:12) should govern? To borrow an observation from David Friedman, those who would have the government tax and redistribute funds to create social justice, further welfare entitlements and human rights, provide health care, education, and so forth, usually envision a government controlled by people very like themselves, who will use the force of government to pursue the one true good society.[15] But, this is simply an appeal to power. If others do not share one's convictions regarding the good, or the good life, they may experience one's attempts to act charitably, to do good, as acts of violence. The Golden Rule can be and has been the basis for the coercive imposition of particular concrete understandings of the good life. In short, without the justification of a canonical moral framework, the suggestion "to treat others as you would wish to be treated" is at best empty and at worst tyrannical.

10.4.3 Assumption 3: What is the Goal of Charity?

Assume for the moment that persons are in fact morally obligated to give to charity out of their surplus. Why would persons be further obliged to discharge this duty in the "most efficacious" manner? Insofar as the goal of charity is solely to address the welfare needs of others, e.g., food, education, health care, then it is seems plausible to say that it would be *useful* or perhaps *prudent* to do so efficiently and effectively, all things considered; but why would efficaciousness necessarily rise to the level of a moral obligation? The plausibility of this additional moral obligation will in part depend on the underlying goals of charity as well as on a presumed ranking of appropriate outcomes.

If the central goal of charity is other than the simple efficient fulfilling of welfare needs necessary for human flourishing, then providing charity in the most efficacious manner to support the welfare-based needs of the impoverished may not be morally obligatory.

Consider some alternatives:

(1) If the goal of charity is to teach traditional Christian spiritual humility—A is humbled by realizing the needs of others and by having to give to help others, and B is humbled by having to ask for help—then the most efficacious welfare distribution solution for welfare needs might not achieve the desired spiritual results. One might prefer less efficient solutions, such as anonymous coordination through one's parish priest, so that there is someone who ensures that the spiritual goals of charity for both giver and receiver are met. As a result, many may appreciate Boyle's characterization of redistributive taxation and social justice as helping to fulfill one's Christian charitable duties as fundamentally misguided. Christian spiritual growth is not a goal that large scale organizations such as the United Way, Planned Parenthood, much less the U.S. or Canadian governments, are capable of producing.

(2) Perhaps the role of charity is to reinforce the good of the traditional family. Here charity will need to be personal and individual, rather than conducted through social organizations or state bureaucracies. There is remarkable evidence that the modern welfare state has played a significant role in the destruction of the traditional family. Amongst the criticisms levied against the now infamous welfare program Aid to Families with Dependent Children was that it encouraged single mothers to have children.[16] Once elderly or needy family members receive significant welfare assistance from the state, there are direct financial incentives for other family members no longer to care for the needy or elderly (Daniels, 1988). Families no longer have a need to live together or to find ways to get along with each other. The redistributive mechanisms of the modern social welfare state make it possible for each member to be financially independent even in infirmity and need. Recognizing the significance of this challenge, current public policy debates in China, for instance, often regard how best to support the traditional family as the core moral and social unit, while still providing long term health care, housing and food for the elderly. Their proposed solutions tend to seek creative ways to support the role of the family in taking care of the elderly (see Fan, 1999; Fan and Tao, 2004; Tao, 1999). For example, tax reductions for adult children who take care of their elderly family members provide incentives to be charitable towards one's parents and, as a result, help to bind the family together.

(3) If the goal of charity is to enrich the personal and social ties of religious or cultural communities, then charity will need to be coordinated through a particular religious community, such as an Orthodox Christian parish, or

through a particular cultural organization, such as a university alumni association. Whether such charity ought to be provided in the "most efficacious" manner to meet welfare needs will necessarily depend on the underlying goals of the religious or cultural community, which may vary considerably. Alumni associations frequently encourage business networking among fellow alumni so as to encourage graduates of the same university to hire each other, even if the fellow alumnus is not necessarily the best job candidate. Minority groups and religions often encourage their members to hire from within the minority group, or from within the same religion or parish. "Give the job to a brother," one might hear the parish priest say, when members of the parish council are considering whom to hire to work in their various businesses. Even in market transactions, one can at times find hidden charitable interests. One might think, for example, "I will purchase my 20-year old Laphroaig from store A, rather than from store B, because the owners of store A give regularly to support my church."

(4) Charity might be about bringing neighborhoods together as communities and enriching the lives of neighbors as friends. While charity is still about meeting the needs of others, here the focus is on the ways in which it brings people together in support of common projects. For example, one neighbor might help another trim his trees—thereby beautifying the neighborhood and making it safer. Another might edit a neighbor's dissertation, learning together and becoming friends. Entire neighborhoods might come together to clean the park, or to create a neighborhood watch program. Derek Parfit and others have decried as irrational what they term "within a mile altruism," i.e., that persons are more concerned with being altruistic towards those who live close to home, and are less concerned with those who live farther away, even if those who live further away are in the same or greater need (Parfit, 1986). However, if the goal of charity is to bring neighborhoods and friends together and to enrich their lives, rather than simply to meet welfare needs, such an attitude is far from irrational.

In general secular terms, the market can be seen as the social space for advertising diverse understandings of charitable obligations and for embodying a wide variety of charity-based goals. Because of the diffusion of social and political authority the market defines social space for peaceable consensual human interaction, even among those who do not share the same background moral or religious beliefs and commitments. Here morality exists as the marketplace of moral ideas and moral understandings within which each peaceably pursues his own ends, including charitable interests, without necessarily sharing a common, content-full moral vision or concrete view of justice (Engelhardt, 1996). While it is often perceived solely in terms of the pursuit of profit, the market does not preclude altruistic action—indeed, in the marketplace of charities, the market directly encourages altruism.

Boyle is correct that churches and other voluntary charitable organizations should play a significant role in creating health care, education, and other welfare resources for the poor. However, there are a wide diversity of churches and charitably based organizations with a significant variety of interests. The market secures the possibility for such moral diversity within particular countries and cities, as well as for the emergence of a worldwide network of non-geographically based communities with their own understandings of moral probity, bioethics, appropriate health care, and charity-based goals. In itself, the market will not resolve such moral diversity, nor will it celebrate such diversity; rather, it leaves such decisions to free and consenting persons. Respecting the freedom of persons to interact with free and consenting others defaults to protecting liberties of association, contract, conscience, and religion, and thereby to protecting the possibility of substantial moral diversity, including divergent incommensurable instantiations of the good life, expressed through divergent incommensurable charity-based goals.

Market transactions and contractual relationships draw moral authority from the consent of the participants to be bound by their agreements. The parties to the transactions themselves freely convey authority to the enforcement of the specified conditions. The actual agreement of actual persons creates and thereby limits the moral authority to interfere in the free interaction of consenting persons—including their agreements to engage in charity to pursue various interests and goals. Here moral authority is not drawn from assertions of so-called "moral consensus," ideal theories of rational action, or even deep moral intuitions regarding consequences, human rights, cardinal moral concerns, or basic human goods. Collaborators need not agree regarding the background ranking of values or moral principles, cultural or religious assumptions, or claims regarding basic goods and the requirements of practical reason; they need only affirm the content of their agreement. Nor must one presume a particular value standard or canonical order—just the recognition that collaboration is possible through agreement. Agreement or permission is the ground of the moral justification of such collaboration. Here, one does not affirm the market—even as the market of ideas and competing moralities—or its outcomes as good in themselves; rather the idea of the market is the creation of social space for unencumbered personal interaction. The market is simply the result of respecting the moral authority of persons over themselves and their private property.

10.5 Conclusion—Charity as Spiritual Therapy

Boyle is correct in this sense: persons have substantial charitable obligations. However, in my view, he has mischaracterized the origin and foundational goals of charity. The origin of charity is love of God. All that we have is from God and all that we have is owed to God. The central goal of Christian charity is not social justice—we should not attempt to create the Kingdom of God on earth—nor is it, in a strict sense, to alleviate the welfare needs of others, although it may alleviate such needs; rather the role of charity is to show love to others, to show love for God, and to learn humility and obedience to God.

Here one might think of St. Theodoros the Great Ascetic (c. A.D. seventh century), who notes that care of the poor is done out of a duty of hospitality; i.e., love for others:

> If you have renounced worldly cares and undertaken the ascetic struggle you should not desire to have wealth for distribution to the poor. For this is another trick of the devil who

> arouses self-esteem in you so as to fill your intellect with worry and restlessness. Even if you have only bread or water, with these you can still meet the dues of hospitality. Even if you do not have these, but simply make the stranger welcome and offer him a word of encouragement, you will not be failing in hospitality. Think of the widow mentioned in the Gospel by our Lord: with two mites she surpassed the generous gifts of the wealthy (Mark 12, 42–44) (Theodoros, 1981, 23).

Or, consider St. Maximos the Confessor (A.D. 580–662) and St. John of Damaskas (A.D. c. 675-c.749), both of whom urge us to appreciate charity as a therapy to sin. St. Maximos writes: "Almsgiving heals the soul's incensive power; fasting withers sensual desire; prayer purifies the intellect and prepares it for the contemplation of created beings. For the Lord has given us commandments which correspond to the powers of the soul" (Maximos, 1981, 61–62). Similarly, St. John writes:

> The sins of the desiring aspect are gluttony, greed, drunkenness, unchastity, adultery, uncleanliness, licentiousness, love of material things, and the desire for empty glory, gold, wealth and the pleasures of the flesh. These are cured through fasting, self-control, hardship, a total shedding of possessions and their distribution to the poor, desire for the imperishable blessings held in store, longing for the kingdom of God, and aspiration for divine sonship (John, 1981, 337).

In each case, the focus of charity is to change one's heart, to learn to love others, rather than oneself, and to humble oneself before God, Himself. Its final goal is salvation.

State-based coercive taxation is incapable of aiming persons rightly in the struggle towards God and salvation. Moreover, as I have argued, each of the core natural law assumptions is seriously questionable, if not straightforwardly question-begging. Governmental taxation for the creation and promotion of welfare entitlement rights is an unauthorized extension of governmental coercive power and is thereby illicit. It corrupts individuals into supporting goals and moral viewpoints that are frequently openly hostile to those which are Traditional Christian—such as taxpayer financed abortion as a welfare entitlement. It corrupts the fundamental role of Christian charity—seeking social justice rather than the struggle towards salvation. In terms of general secular morality, citizens are not morally or rationally required to understand themselves as being obliged to pay taxes to support welfare entitlements; rather, paying taxes is, at best, a prudential response to stave off the possibility of yet more illegitimate governmental coercion, such as arrest, fine, and wage garnishment. Honestly paying taxes to support illegitimate welfare entitlements can be seen on analogy with prudentially paying protection money.[17] This conclusion raises further interesting philosophical questions regarding the moral propriety of tax evasion, but the exploration of such questions reach beyond the philosophical issues at the heart of my arguments in this paper.

Even though I presume that Professor Boyle will disagree with at least some of my conclusions, and will have a word or so to say in response, I also presume that he would be at peace exploring the arguments with me at some point, perhaps over an appropriate glass or two of Bushmills fine Irish whiskey.

Notes

1. Boyle and his collaborators describe the categories of basic goods as follows: "As *animate*, human persons are living organic substances. Life itself—its maintenance and transmission—health, and safety are one form of basic human good. . . .As *rational*, human beings can know reality and appreciate beauty and whatever intensely engages their capacities to know and feel, and to integrate the two. Knowledge and aesthetic experience are another category of basic good. As simultaneously *rational* and *animal*, human persons can transform the natural world by using realities, beginning with their own bodily selves, to express meanings and/or serve purposes within human cultures. Such bestowing of meaning and value can be realized in diverse degrees; its fullness is another category of basic good: excellence in work or play. . . . As *agents through deliberation and choice*, they can strive to avoid or overcome various forms of conflict and alienation, and can seek after various forms of harmony, integration, and community (fellowship). . . . Most obvious among the basic human goods of this relational dimension are various forms of harmony between and among persons and groups of persons: friendship, peace, fraternity, and so on" (Finnis, Boyle, and Grisez, 1987, 279–280)
2. Among the leading causes of death for 2003, the Center for Disease Control (CDC) lists 65,163 deaths due to influenza and pneumonia. The CDC estimates that between 5 and 20% of U.S. residents suffer from influenza each year, more than 200,000 require hospitalization, with more than 36,000 Americans dying yearly from influenza complications (CDC, 2007). Note, the U.S. does not forcibly inoculate for influenza and pneumonia, rather it stipulates classes of high risk persons, and recommends that they and others should likely seek immunization.
3. As John Locke pointed out, governments that use the property and persons of their citizens without permission are conceptually no different than thieves: "Wherever law ends, tyranny begins, if the law be transgressed to another's harm; and whosoever in authority exceeds the power given him by the law, and makes use of the force he has under his command to compass that upon the subject which the laws allows not, ceases in that to be a magistrate, and acting without authority may be opposed, as any other man who by force invades the right of another" (1980 [1690], 103 §202).
4. Pluralistic moral casuistry typically considers rights to be one among other moral appeals to be considered in reasoning about particular cases. See Brody (1988); Cherry and Iltis (2007).
5. "In contrast to incorporating rights into the end state to be achieved, one might place them as side constraints upon the actions to be done: don't violate constraints C. The rights of others determine the constraints upon your actions" (Nozick, 1974, 29).
6. "This is to say, *if* natural rights are understood as defining for each person a basic sphere of moral jurisdiction within which each rightholder may do as he sees fit, then it seems obvious that no feature or aspect of a person's rights can

block the moral permissibility of chosen actions within his protected domain" (Mack, 1999, 143). Property rights are often described in terms of a sphere of moral jurisdiction: "Property rights thus build a sphere of privacy around a person, within which he or she is free to act without anyone else interfering: a realm of personal dominion where people may make their own choices" (Sandefur, 2006, 22).

7. "When we call anything a person's right, we mean that he has a valid claim on society to protect him in the possession of it, either by the force or law, or by that of education and opinion. If he has what we consider a sufficient claim, on whatever account, to have something guaranteed to him by society, we say that he has a right to it" (Mill, 1979 [1861], 52). Mill's justification for such a right is only that it maximizes the good: "...I can give no other reason than general utility" (Mill, 1979 [1861], 52). For consideration of wealth maximization see Posner (2002).
8. Consider, for example, the United Nation's Universal Declaration on Human Rights: no arguments are provided in support of the substantial list of rights, nor does the United Nations possess the funds or coercive power to secure the rights; rather, each right is straightforwardly announced as a rhetorical method of setting out an aggressive agenda for national and international social reform.
9. Genesis chapter nine lays out what the covenant with Noah requires of Gentiles, which includes, for example, capital punishment for murder: "But you must not eat meat that has its lifeblood still in it. And for your lifeblood I will surely demand an accounting. I will demand an accounting from every animal. And from each man, too, I will demand an accounting for the life of his fellow man. Whoever sheds the blood of man, by man shall his blood be shed; for in the image of God has God made man" (Genesis 9: 4–6).
10. Such contracts are currently illegal in most jurisdictions worldwide. For an analysis of why they ought to be legally permissible and enforceable see Cherry (2005).
11. Jutta M. Joachim documents the real anger directed at the Vatican for blocking various proposals at the International Conference on Population and Development in Cairo, 1994, that focused on easy access to abortion and contraception. "To prevent the inclusion of reproductive rights and reproductive health on the conference agenda, the Holy See together with Catholic countries responded by entering into an alliance with fundamentalist Islamic countries, such as Iran, Libya, and the Sudan, despite their ideological differences. The alliance successfully blocked discussion on chapters 7 and 8, devoted to women's reproductive rights and health, by making ample use of UN rules and procedures, such as frequent oral interventions and the demand for brackets, indicating their disagreement....While the opposition of the Holy See was a significant obstacle, it remains unclear why the Women's Alliance was unable to overcome it" (2007, 150–151).
12. According to Health Canada, in British Columbia, Alberta, Ontario, Newfoundland, and most facilities in Quebec, hospital and clinic fees for

abortion are paid by the province. In the other provinces and territories, only abortions performed in hospitals are funded. See www.hc-sc.ga.ca.

13. As Sandefur notes: "Property rights minimize the degree to which people may use *force* against each other and require them to deal with one another on the basis of mutual respect, deliberation, and cooperation. This increases their freedom" (2006, 22).
14. Chief Justice Marshall argued: "That the power to tax involves the power to destroy; that the power to destroy may defeat and render useless the power to create; that there is a plain repugnance, in conferring on one government a power to control the constitutional measures of another, which other, with respect to those very measures, is declared to be supreme over that which exerts the control, are propositions not to be denied. But all inconsistencies are to be reconciled by the magic of the word CONFIDENCE. Taxation, it is said, does not necessarily and unavoidably destroy. To carry it to the excess of destruction would be an abuse, to presume which, would banish that confidence which is essential to all government" *McCulloch v. Maryland* 17 U.S. 316, 327 (1819).
15. "Since most socialists imagine a socialist government to be controlled by people very like themselves, they imagine that it will pursue the true good—the one that they, imperfectly, perceive. That is surely better than a chaotic system in which all sorts of people other than the socialists perceive all sorts of other goods and waste valuable resources chasing them. People who dream about a socialist society rarely consider the possibility that some of those other people may succeed in imposing their ends on the dreamer, instead of the other way around" (Friedman, 1989, 18).
16. As Francis Fukuyama encapsulates the political and social history: "Moynihan's critique was extended by Charles Murray, who pointed to the unanticipated consequences of welfare programs like Aid to Families with Dependent Children (AFDC), which encouraged out-of-wedlock births and contributed to the culture of poverty. This critique of AFDC led ultimately to its abolition under the Personal Responsibility and Work Opportunity Reconciliation Act of 1996, initiated by the Republican Congress and signed by President Bill Clinton" (2006, 20). For the original Moynihan report, see Moynihan (1965); see also Wilson (1988).
17. "Greedy capitalists get money by trade. Good liberals steal it" (Friedman, 1989, 85).

References

Boyle, J. 1999. Personal responsibility and freedom in health care: A contemporary natural law perspective. In *Persons and their bodies: Rights, responsibilities, relationships*, ed. M.J. Cherry, 111–142. Dordrecht: Kluwer.

Boyle, J. 2001. Fairness in holdings: A natural law account of property and welfare rights. *Social Philosophy and Policy* 18:206–226.

Boyle, J. 2006. The bioethics of global biomedicine: a natural law reflection. In *Global bioethics: The collapse of consensus*, ed. H.T. Engelhardt, Jr., 300–334. Boston: M&M Scrivener.

Brody, B.A. and H.T. Engelhardt, Jr. 1987. *Bioethics: Readings and cases*. Englewood Cliffs, NJ: Prentice Hall.

Brody, B.A. 1988. *Life and death decision making*. New York, NY: Oxford University Press.

Center for Disease Control 2007. Influenza (Flu): Questions & answers: The disease. Available at, www.cdc.gov/flu/about/qa/disease.htm. Accessed February 20, 2007.

Cherry, M. 2005. *Kidney for sale by owner: Human organs, transplantation, and the market*. Washington, DC: Georgetown University Press.

Cherry, M. and A. Iltis 2007. *Pluralistic casuistry: Moral arguments, economic realities, and political theory*. Dordrecht: Springer.

Donagan, A. 1977. *The theory of morality*. Chicago, IL: The University of Chicago Press.

Daniels, N. (1988). *Am I my parents' keeper? An essay on justice between the young and the old*. New York, NY: Oxford University Press.

Engelhardt, H.T. Jr. 1996. *The foundations of bioethics, 2nd edition*. New York, NY: Oxford University Press.

Fan, R. 1999. Freedom, responsibility, and care: Hong Kong's health care reform. *The Journal of Medicine and Philosophy* 24:555–570.

Fan, R. and J. Tao 2004. Consent to medical treatment: The complex interplay of patients, families, and physicians. *The Journal of Medicine and Philosophy* 29:139–148.

Finnis, J., J. Boyle and G. Grisez 1987. *Nuclear deterrence, morality, and realism*. New York, NY: Oxford University Press.

Friedman, D. 1989. *The machinery of freedom: Guide to radical capitalism*. Chicago, IL: Open Court.

Fukuyama, F. 2006. *America at the crossroads: Democracy, power, and the neoconservative legacy*. New Haven, CT: Yale University Press.

Joachim, J.M. 2007. *Agenda setting, The UN, and NGOs: Gender violence and reproductive rights*. Washington, DC: Georgetown University Press.

John of Damaskos, St. (1981). On the virtues and the vices. In *The Philokalia*, eds. G.E.H. Palmer, P. Sherrard, and K. Ware, 334–342. (compilers St. Nikodimos and St. Makarios). London: Faber and Faber.

Locke, J. 1980 [1690]. *Second treatise on government* ed. C.B. Macpherson. Indianapolis, IN: Hackett.

Mack, E. 1999. The alienability of Lockean natural rights. In *Persons and their bodies: Rights, responsibilities, relationships*, ed. M.J. Cherry, 143–176. Dordrecht: Kluwer.

Maximos the Confessor, St. 1981. Four hundred texts on love. In *The philokalia*, eds. G.E.H. Palmer, P. Sherrard, and K. Ware, 52–113. (compilers St. Nikodimos and St. Makarios). London: Faber and Faber.

Mill, J.S. 1979 [1861]. In *Utilitarianism* ed. G. Sher. Indianapolis: Hackett.

Moynihan, D.P. 1965. *The Negro family: A case for national action*. Washington, DC: U.S. Department of Labor.

Nozick, R. 1974. *Anarchy, state, and Utopia*. New York, NY: Basic Books.

Parfit, D. 1986. *Reasons and persons*. New York, NY: Oxford University Press.

Posner, R.A. 2002. *Economic analysis of law*. New York, NY: Aspen.

Sandefur, T. 2006. *Cornerstone of liberty: Property rights in 21st century America*.Washington, DC: Cato.

Schmidt, K. 2002. Stabilizing or changing identity? The ethical problem of sex reassignment surgery as a conflict among the individual, community, and society. In *Cross-cultural perspectives on the (im)possibility of global bioethics* ed. J. Tao , 237–264. Dordrecht: Kluwer.

Tao, J. 1999. Does it really care? The Harvard report on health care reform for Hong Kong. *The Journal of Medicine and Philosophy* 24:571–590.

Theodoros the Great Ascetic, St. 1981. A century of spiritual texts. In *The Philokalia*, eds. G.E.H. Palmer, P. Sherrard, and K. Ware, 14–47 (compilers St. Nikodimos and St. Makarios). London: Faber and Faber.

United Nations. 1996. *HIV/AIDS and human rights*. Geneva: United Nations.

Wilson, W.J. 1988. *The truly disadvantaged: The inner city, the underclass, and public policy.* Chicago, IL: University of Chicago Press.

Chapter 11
Natural Law, Property, and Welfare Rights

Andrew Lustig

It is a distinct pleasure to contribute to this festschrift volume honoring Joseph Boyle. I have admired, and learned much from, the wide range Dr. Boyle's scholarship—his contributions to so-called "revised" natural law epistemology, his analysis of the moral status of nuclear deterrence as a strategy, and his perceptive discussions of particular issues in bioethics and medical ethics.

In these remarks, I will focus on perhaps a less well-known topic in Boyle's corpus, viz., his natural law analysis of the conceptual links between the justification of property and the nature and scope of welfare rights as political entitlements. While his arguments have broad implications for particular positive rights he has also discussed, such as the right to health care (e.g., Boyle, 1996), I will concentrate on his more general account, because it provides the moral underpinning and larger context for justifying specific entitlements.

Among Boyle's works, his essay, "Fairness in Holdings: A Natural Law Account of Property and Welfare Rights" (Boyle, 2001, 206–226), offers the most detailed natural-law analysis of the relations between property and welfare rights. The structure of that article is worth elaborating in some detail, because his natural law defense of the institution of private property is *both* individual and communal in its implications, with features that are distinct from other approaches to positive rights that are not based on natural law grounds.

Boyle understands welfare rights as "the grounding of duties in the interest or welfare of other people" (Boyle, 2001, 206). Here, he seconds the perspective offered earlier by John Finnis: welfare rights, as individual or group entitlements, are particular instantiations of positive moral duties incumbent on individuals, groups, and the state to meet certain basic needs (Finnis, 1980). In Boyle's words, welfare rights are "claims ... to provide a fair share of those instrumentalities or empowerments important for living good human lives" (Boyle, 2001, 206). But in order for such *moral* claims to be implemented in modern societies, Boyle also argues that they must be instantiated as *political* rights: viz., "legally recognized and established claims by members of a political society on that society as a community,

A. Lustig (✉)
Davidson College, Davidson, NC, USA
e-mail: anlustig@davidson.edu

C. Tollefsen (ed.), *Bioethics with Liberty and Justice*, Philosophy and Medicine 110,
DOI 10.1007/978-90-481-9791-0_11,

whose leaders and members have duties based upon the interests of those who have the rights in question" (Boyle, 2001, 206). In addition, in order for such political entitlements to be implemented, Boyle argues that some form of redistributive taxation will be a practical corollary of their implementation, because "[a] political society cannot guarantee such support for people in need without some form of taxation" (Boyle, 2001, 207).

The issue of taxation, of course, immediately poses questions for commentators predisposed to view such welfare claims with skepticism, and critics of Boyle's defense of such taxation for welfare purposes can be found both within and without the Catholic commentariat. Boyle himself appreciates the need for careful argument on behalf of his claims, because the "[t]he justification of political welfare rights ... must be grounded in moral considerations sufficiently weighty to justify some redistributive taxation" (Boyle, 2001, 207).

To be sure, modern Catholic social teaching, especially the encyclical tradition since the papacy of John XXIII, has affirmed that welfare rights are the material concomitants of human dignity. Nonetheless, Boyle expresses reservations about the usefulness of such affirmations about natural links between welfare rights and human dignity, because much of the time, the association claimed reads more in the nature of assertion than carefully grounded argument. In Boyle's opinion, such statements, whatever their prophetic thrust, fail to provide a sufficient account of how welfare rights are to be philosophically justified. As he comments, "[T]here is a lack of developed analysis showing how welfare rights emerge from more fundamental moral principles, and there are perhaps even some significant tensions between a characteristic natural law affirmation of property rights and a willingness to allow redistributive taxation" (Boyle, 2001, 208). Indeed, he suggests, "one might be led to suppose that such talk on the part of Catholic exponents of natural law is overstated, or simply an arresting way of restating traditionally held obligations that are really social rather than political" (Boyle, 2001, 207).

The distinction here between social and political rights is an important one in the Catholic tradition, especially in its Scholastic roots. According to Boyle, natural law theorists since the time of Aquinas have emphasized a common duty to assist the needy but have depicted that duty as one initially incumbent on individuals or small groups rather than on society at large. Such an understanding is one aspect of the so-called principle of subsidiarity, which, as enunciated formally by Pius XI, finds both intrinsic and instrumental value in lodging such obligations at the lowest level of appropriate responsibility, that of voluntary associations whenever possible. Given his reservations, Boyle proceeds with a sequential argument that he believes will provide the philosophical bridgework lacking in much of recent Catholic welfare rights discussion. He first rehearses Aquinas's justification of property rights and their appropriate limits. He then considers the compatibility of that justification of private property with warrants for welfare rights. Finally, he considers whether the moral logic of Thomas's account, as applied to the circumstances of modern social life in developed economies, implies that social obligations "are properly coordinated by political authority" (Boyle, 2001, 208). In what follows, I will review the premises at work in Boyle's argument, especially his analysis of property holding and the common good as developed in Aquinas. I will then look in some detail at Boyle's natural-law defense of welfare rights in modern societies, and will compare and contrast his analysis with other recent discussions.

11.1 Property Rights and the Common Good in Aquinas

According to Boyle, Thomas's discussion of property, which occurs most fully in his discussion of stealing, initially focuses only on the status of possession as "natural," in the sense that such possession is appropriate or morally justified in light of our rational nature (Aquinas, *ST*, II-II, Q. 66, a. 1). Boyle begins by observing that Thomas does not "say or suggest that things are in some way the common property of all. His analysis does not concern the relationships between humans that arise because of the common human need to use things; instead, it concerns the relationships between humans as such, on the one hand, and God and things on the other" (Boyle, 2001, 209).

Boyle's textual point here is literally accurate, although I believe that its emphasis on the sequence of Thomas's discussion may be a bit misleading. Boyle is surely correct that our ability to appropriate and develop resources is, in the first instance, grounded in our rational nature as creatures made in the image of God ("humans as such"). Yet that philosophical point is, I suggest, not isolable from its moral implications regarding common use in the order of practical reason. Rather, the appropriateness of possessing external things as "natural" for Thomas must be situated, from the first, within the context of his characteristic teleology and the central role that *recta ratio* plays in his metaphysics. And here, John Finnis's analysis of Thomas's discussion as foundationally social appears to more accurately reflect Thomas's moral logic. Finnis argues as follows,

> Though Aquinas has no formal treatment of ways in which property rights are acquired, he recognizes 'taking possession' [occupation] as a legitimate mode of acquisition. But he does so on the basis that this appropriation of "what from the outset was common" (i.e., available to all) is precisely a means whereby the owner to whom things have been appropriated "gets things ready for" and shares them with, other people [*aliis communicat*].
>
> Indeed, all the justifications for appropriation of resources to particular owners . . . are based on "general" justice, i.e., on the advantages which such appropriation is likely to bring to all members of the community . . . the ultimate use of resources in consumption must remain fundamentally "common" (unappropriated). Any appropriation of resources to particular owners is always subject to this reservation (Finnis, 1998, 189–90).

Why is this foundationally social emphasis important? From the first, in the order of practical reflection, practical reason discerns three natural law directives as *per se nota* truths (truths that are self-evident but underived): (1) the preservation of mankind; (2) the preservation of society; and (3) the worship of God (Aquinas, *ST*, I-II, Q. 94, a. 2). These are *primary* directives or precepts of natural law because we rationally apprehend them as goods to be pursued. Property, understood in its most general sense as appropriation, whether private or communal, is available to humans because they possess and exercise right reason; humans are therefore capable of exercising *rational* dominion because they share an intellectual nature. In light of the foundationally social nature of the first two directives above, and because God has given mankind the earth in common, the primary relationship between humans as rational creatures and material resources for Thomas is that of *use*. Man is empowered, in the words of one commentator, "in virtue of his reason and will, to make use of things for his own benefit" (Parel, 1979, 97).

Boyle, although initially addressing the philosophical basis of man's relationship to material things in general terms, does move quickly to consider the "[q]uestions about the relationships among humans that arise because of their common need to use things," an issue addressed explicitly by Aquinas in his discussion "of whether it is permissible for anyone to possess something as his or her own (*quasi proprium*)" (Boyle, 2001, 209). Boyle distinguishes two different aspects of property in Thomas's discussion: *viz.*, "the authority to take care of and distribute things [what Boyle stipulates as 'ownership'] and their actual use" [which, for Boyle as well, is linked to things being held in common] (Boyle, 2001, 209).[1] Thomas's well-known account of the authority to appropriate and distribute (by individuals or groups) is based on several prudential concerns. Thus he says,

> Two things are competent to man in respect to exterior things. One is the power to procure and dispense them, and in this regard, it is lawful for man to possess property. Moreover, this is necessary to human life for three reasons, First, because every man is more careful to procure what is for himself alone than that which is common to many or to all: since each would shirk the labor and leave to another that which concerns the community, as happens when there are a great number of servants. Secondly, because human affairs are conducted in more orderly fashion if each man is charged with taking care of some particular thing himself, whereas there would be confusion if everyone had to look after any one thing indeterminately. Thirdly, because a more peaceful state is ensured if each man is contented with his own. Hence it is to be observed that quarrels arise more frequently where there is no division of things possessed (Aquinas, *ST,* II-II, Q. 66, a. 2).

Notice here, however, that such reasons for having authority to appropriate and distribute things are invoked within the foundationally social perspective revealed in the first two *per se* nota truths quoted earlier. Boyle emphasizes that all resources, in respect to a second aspect—that of their use—"should be held in common" (Boyle, 2001, 210).

I have argued elsewhere that Thomas's warrant for our exercise of natural dominion is expressly theological. God has made the earth, and humans, by virtue of their intellectual nature, have been granted dominion over it (Lustig, 1991, 123). That theological tenet establishes, admittedly in very broad terms, the appropriate limits of human stewardship. As James Tully aptly comments, "The world is properly God's property, so man has no power over its substance" (Tully, 1979, 65). The broader implications of that theological warrant are directly pertinent to the way that we are to understand subsequent judgments about appropriate holdings of property. First, the goods of the earth initially "belong" to humankind in general rather than to individual "owners" (so long as the terms in quotes are understood in moral terms rather than as juridical or political concepts). Second, as we have seen, the nature of property holding is fundamentally social in light of the first two *per se nota* truths I have discussed: access to resources is established, in the order of moral reflection, primarily in terms of common *use* rather than private ownership. Third, the sole discussion of the appropriate end of *surplus* of property that is individually appropriated is social use.

Perhaps something more should be said here about the language Thomas employs in his discussion. Boyle notes that Aquinas uses different terms for property, depending upon the type of "ownership" that he discusses. When speaking of common ownership, he uses the words *dominium* and *possessio*. When referring to private holdings, he uses the word *proprietas*. Indeed, Thomas develops the concept of *dominium* in his discussion of property in analogical fashion; it is tied to the philosophical notion of dominion that reason exercises over acts of mind and will. To this extent, dominion is not to be confused or conflated with juridical notions of property. Rather, as Anthony Parel suggests,

> *Dominium* ... is a broad, indeterminate power that every man has with respect both to internal acts of mind and will as well as to the use of the things of this world. It is an analogical notion. Possession, as distinct from *dominium*, is a specification and an extension of the latter to material things (Parel, 1979, 96).

Nonetheless, because property for Thomas is a material object to be ordered by right reason, *possessio* rather than *proprietas* assumes moral priority in his discussion. Granted, this is not an uncontroversial judgment, because much is made in Catholic social thought about the "naturalness" of private property. But *dominium* as a power appropriately exercised in accordance with right reason, looks first to the *per se nota* natural law directives that are immediately available to practical reason in moral reflection. And given the social focus of the first two such directives—the preservation of mankind and society—*possessio*, rather than *proprietas*, emerges as the primary material expression of *dominium*.

Moreover, in elaborating the meaning of *dominium*, Thomas expands the scope of classical conceptions. According to Richard Tuck, *possessio* and *dominium* were sharply contrasted in classical law: "to possess something was to occupy it and use it but not to have private property rights in it" (Tuck, 1979, 18). During the medieval period, however, the concept of usufructuary *possession* as a use right began to be viewed as properly within the domain of *dominium naturale*, especially by Accursius of Bologna during the second and third decades of the thirteenth century (Tuck, 1979, 16–17). Use rights, rather than being understood as functions of the natural rather than the civil order, were now described as *powers* that could be validly exercised against other persons. Tuck concludes as follows:

> ... the recognition of the category of *dominium utile* was to transform rights theories. For now, *dominium* was taken to be any *ius in re*: any right which could be defended against all other men, and which could be transferred or alienated by its possessor, was a *property* right, and not only rights of total control. The process had begun whereby all of a man's rights, of whatever kind, were to be seen as his property. This obscure thirteenth-century feud had these tremendous consequences: there is a direct line linking Accursius with the

> late medieval rights theorists, and through them with the great seventeenth-century figures (Tuck, 1979, 16).

The development of the concept of *dominium utile*, whereby even a usufructuary exercises dominion of a sort, allows Thomas to extend the category of natural dominion to include *powers* of property. Unlike earlier theorists, who, much in the spirit of Augustine (Schlatter, 1951, 37–39), had argued that natural law generates only usufructuary rights, and that *dominium* is at most a conventional construct of the *ius gentium*, Aquinas, by way of Accursius, can now include *dominium* within the realm of natural law. He takes this important step by distinguishing two senses of natural law:

> Something can be said to be according to the *ius naturale* in two ways. One, if nature inclines us to it: such as not to harm another human being. The other, if nature does not prescribe the opposite: so that we can say that a man is naked under the *ius naturale*, since he received no clothes from nature but invented them himself. In this way, "the common possession of all things and the equal liberty of all" is said to be according to the *ius naturale*: for distinctions between possessions . . . were not the products of nature, but were made by human reason for the advantage of human life (Aquinas, *ST*, I-II, Q. 94, a. 5).

Granted, Thomas's understanding here does not entail an individual "right" as we understand the term. Strictly speaking, *dominium* as a human power of possession remains analogical and is always relative to the absolute ownership that God alone enjoys. Thus, somewhat in contrast to Boyle's initial judgment, I find in Thomas's discussion of the relations that obtain between God, humankind, and material resources *direct* implications for the moral priority of use rights, understood in light of foundationally social natural law directives. That is to say, right reason, which provides the basis for Boyle to describe the appropriate metaphysical relations among God, creatures, and material resources has, I believe, immediate moral implications.

At the same time, as all commentators agree, Thomas indeed justifies private property, including the presence of surplus in circumstances of relative abundance. Given my own emphasis on the foundationally social perspective at work in Aquinas's account, the three largely prudential reasons with which Thomas justifies ownership authority are instructive at two levels: first, for appreciating that nature "does not prescribe the opposite," that is, that private possessions are not against natural law in Thomas's first sense above; and second, that private holdings are both justified and limited by both senses of natural law as expressed above. That is, private holdings are rationally justified by the second sense of natural law above—as appropriately advantageous to human life—but also remain limited by the first sense of natural law above—only if superabundance is not harmful to others in dire need. Right reason, illuminated by foundationally social *per se nota* truths, provides the basis for both justifying and limiting ownership authority.

In broad strokes, then, Thomas's justification of private property emerges as a *via media* between the earlier Roman notion of *dominium* as an absolute power to control and dispose of one's possession and the Platonic defense of communism rejected by Aristotle in his *Politics*. Thomas draws heavily on Aristotle's account of property (Aquinas, *ST*, I-II, Q. 105, a. 2), but he also relies on the discussion of property in common in the Stoic tradition, especially in the writings of Cicero and Seneca. As Richard Schlatter emphasizes, the Stoics taught that men have a "common right to use things produced by nature" (Schlatter, 1951, 25–26). For Cicero and for Seneca 100 years later, all men are by the law of nature equal. In the putative "golden age" when men lived according to the law of nature, equality was the rule and "property was common and used equally by all for the satisfaction of needs" (Schlatter, 1951, 26).

According to Thomas, the establishment of private property follows from the introduction of sin. Thus he says,

> In our present state, a division of possessions is necessary on account of the multiplicity of masters, inasmuch as community of possessions is a source of strife, as the Philosopher says ... In the state of innocence, however, the will of men would have been so ordered that without any danger of strife they would have used in common, according to each one's need, those things of which they were masters—a state of things to be observed even now among many good men (Aquinas, *ST*, I, Q. 98, a. 3).

The last sentence in the above passage, however, reveals a major difference between Thomas's justification of private property and that of many of the Fathers, including Augustine (Schlatter, 1951, 33–46). For Thomas, fully communal living, including property in common, remains a present possibility for those who live according to the counsels of perfection. His own Dominican order, for example, required renunciation of private property. And for the "better" man, according to Thomas, the "charity of Christ... makes things common rather than political friendship" (*Contra Impug.*, 7, in McDonald, 1939, 81). Moreover, Thomas leaves open the possibility that all could live this way, with God's grace; that is, Thomas sees no necessary connection between the fall and the institution of private property.

As we have seen, Thomas offers several largely prudential arguments to justify ownership authority by individuals and groups (Aquinas, *ST,* II-II, Q. 66, a. 2). But he also provides a more explicit argument in natural law terms. On balance, it appears that Thomas deems private property "natural" by what he calls the way of "determining." Thus,

> [C]ommunity of all possessions and universal freedom may be said to be of the natural law because the distinction of possessions and slavery were not brought in by nature, but devised by human reason for the benefit of human life (Aquinas, *ST,* I-II, Q. 94, a. 5).

The "determination" of property in common into private holdings is properly an addition of the law of nations and, accordingly, "natural":

> It is to be concluded that the *ius gentium* is, in a certain way, natural to man, inasmuch as it is in conformity with reason as derived from natural law, by way of inference, which is not far removed from principle: wherefore men are easily agreed upon it (Aquinas, *ST*, I-II, Q. 95, a. 4)

The law of nations, however, does not prescribe any particular division of property. Accordingly, specific property arrangements are the appropriate subject of civil or positive law. The ultimate warrant for all property arrangements, however, remains Thomas's primary natural law appeal to the common good: "whatever certain people have in superabundance is due, by natural law, to the purpose of succoring the poor" (Aquinas, *ST*, II-II, Q. 57, a. 1).

In some respects, Thomas's view shares affinities with some current notions of zero-sum economics. In his mind, surplus and want are directly related to one another: "With material possessions, it is impossible for one man to enjoy superabundance without someone else suffering want, since the resources of this world cannot be possessed by many at one time" (Aquinas, *ST*, II-II, Q. 118, a. 1). Thus, even with the establishment of private property based on positive law, *dominium utile* would appear to retain its moral priority. The classical notion of dominion as absolute control allowed one to dispose of property without reference to social need. By contrast, Thomas discusses the obligation of the rich to aid the poor as a matter of *justice* rather than of charity. The following passage is characteristic:

> [M]an ought to possess external things, not as his own, but as common, so that, to wit, he is ready to communicate them to others in their need . . . A rich man does not act unlawfully if he anticipates someone in taking possession of something which at first was common property, and gives others a share: but he sins if he excludes others indiscriminately from using it . . . When Ambrose says: *Let no man call his own that which is common*, he is speaking of ownership as regards use (Aquinas, *ST*, II-II, Q. 66, a. 2).

In times when some are in dire need, therefore, all property held in superabundance is rightly considered common:

> Now, according to the natural order instituted by divine providence, material goods are provided for the satisfaction of human needs. Therefore, the division and appropriation of property, which proceeds from human law, must not hinder the satisfaction of man's necessity from such goods. Equally, whatever a man has in superabundance, is owed, of natural right, to the poor for their sustenance. So Ambrose says, and it is also to be found in the *Decretum Gratiani*: "The bread which you withhold belongs to the hungry; the clothing you shut away, to the naked; and the money you bury in the earth is the redemption and freedom of the penniless" (Aquinas, *ST*, II-II, Q. 66, a. 7).

Because Thomas emphasizes *dominium utile*, need emerges here as his substantive criterion of justice. Granted, Thomas distinguishes two different cases of need—one that of a person in "urgent and evident necessity," the other of someone in lesser need (Aquinas, *ST*, II-II, Q. 66, a. 7), and only the first of these cases is relevant to the discussion of welfare rights to which we will turn shortly. There, the man in "urgent and evident necessity" who takes what he needs from another's superabundance does not, properly speaking, steal: necessary provisions are rightfully his by virtue of his need. His dire need, after all, is the shared basis for his right to the means of his subsistence and the duty incumbent on the property holder with more than he requires to meet his vocational obligations to self and specific others.

11.2 Private Property and the Order of Charity

It is important to emphasize that the above criterion of "urgent and evident necessity" provides an important limit to an otherwise unduly expansive notion of common use that some might seek to find in Aquinas's thought. As we have already noted, Thomas adduces pragmatic reasons for establishing private property

(Aquinas, *ST* II-II, Q. 66, a. 2). Boyle describes these as considerations of "fair and rational utility" (Boyle, 2004, 216) in service to the common good ordinarily served by the authority of particular ownership. Such considerations are overridden only in circumstances of extreme and immediate need, and justified in such exceptional cases according to the same foundationally social directives that support private property in ordinary circumstances.

Yet Thomas's own justification of private holdings should not be understood only in light of such prudential considerations. More broadly, his general account of property should be viewed against the backdrop of his theory of virtue. As one commentator puts it, "the balance between the inner force of virtue and the outer force of coercion of just laws is the basis for a humane organization of property" (Parel, 1979, 89). Boyle, in a separate and more general analysis of duties to others in Roman Catholic thought, clarifies and extends his discussion of property by linking the appropriateness of private property and ownership authority to Catholicism's traditional emphasis on an order of charity within which individuals live out the distinctive commitments of their vocations. As Boyle notes,

> ... our willingness to act for the interests of some, such as family members and others to whom we are bound by special ties, should take priority over our willingness to act for others, such as strangers (Aquinas, *ST*, II-II, Q. 26, aa. 6–7). Besides the fact that people are often in a unique position to help those with whom they are bound by special ties of affection and commitment, the interpersonal relationships that develop among those having close human ties are themselves intrinsically valuable and so an irreducible part of the underlying communitarian goal of Christian life (Boyle, 1994, 82).

Moving beyond Thomas's own justification of private property,[2] Boyle commends this individual discretion as not simply a socially useful aspect of property arrangements, but as an important means to the life of virtue in living one's vocation. Thus Boyle says,

> Without some discretion over property and other aspects of life, a person's ability to maintain life and health, to work creatively, to respond to others generously, to worship God appropriately, and to create a good character are seriously limited ... The natural law account of property includes the idea that ownership not only facilitates the proper use of things and promotes harmony, but also provides a form of discretion that is a component of many dimensions of a good life, including virtuous living itself (Boyle, 2001, 214).

Nonetheless, even with Boyle's emphasis here on the relations between discretion over property and Thomas's perfectionist account of virtue, no one should equate such theologically constructed discretion with the autonomy at work in secular discussions of property rights that do not rely on natural law. Here I emphasize again Thomas's own justification of private holdings. He does not justify private property as a primary conclusion of natural law, but by what we have discussed above as the way of "determining," i.e., as a secondary addition of the *ius gentium*. Thus, the primary moral logic at work in Thomas remains "community of possession" (Aquinas, *ST*, I-II, Q. 94, a. 5) and any and all specific property arrangements are subject to civil or positive law "according as each city [!] determines what is suited to itself" (Aquinas, *ST*, I-II, Q. 95, a. 4). Although discretion may be important once specific property arrangements are in place, the moral priority of property in common (and its vocabulary of *dominium utile*) rather than private property (and its corresponding vocabulary of *proprietas*) remain clear, in light of primary natural law directives.

As we have seen, Thomas is not some proto-theorist of individual rights, with private property somehow sacrosanct. Indeed, Thomas is not a theorist of political "rights" in the modern sense at all. Private property is not, for him, an individual claim legitimated by some analogue to the state of nature, but an arrangement "determined" as a secondary conclusion of the natural law, according to the positive or civil law (Aquinas, *ST,* I-II, Q. 95, a. 4). As Thomas makes clear in "On Kingship," it is the ruler, finally, who is charged to distribute and regulate property for the common good. Only within and secondary to appropriate communal authority do notions of individual discretion make sense. In that vein, interpreters of Thomas who find no basis in his thought for recent Catholic social teaching affirming welfare entitlements appear to ignore the priority of the common good in his overall discussion.

11.3 Assisting the Needy in the Modern World: Moral Duties and Political Rights

Thomas' account of property is situated within his robust understanding of virtue, with virtue itself lived out according to dictates of the common good. His discussion focuses on the duty to assist the needy as "a responsibility of individual owners, one generally to be carried out at their discretion even when the requirement is a strict obligation" (Boyle, 2001, 216). Put another way, a person's moral duty to assist the needy is viewed as an imperfect obligation; while a genuine duty, the duty-bearer retains discretion in discharging his or her obligation (in the context of our discussion, discretion in the use of his property for the common good). Thus, in Thomas's time, as in our own, while the general duty to assist the needy is morally binding on natural law grounds, the ways one fulfills it may vary widely. For example, in current circumstances, one may donate one's time and efforts to soup kitchens, to meals-on-wheels programs, to the Saint Vincent de Paul Society, to nursing home visits, to respite care for shut-ins and their families, and so forth. Alternatively (or additionally), one may fulfill one's duty by sharing material resources—for example, supporting particular charities and institutions—the local hospital, Oxfam, Catholic Charities, Worldvision, Bread for the World, and so forth. Biblically, the obligation of tithing was meant to capture both the reality of and some sense of the practical limit to such duties.

This sense of practical limits, as we have seen, is justified in Aquinas, and in Catholic thought more generally, by the order of charity by which individuals live out the role-related obligations of their respective vocations. Nonetheless, what emerges in the later Catholic tradition is the recognition that, as modern society has become more complex, the duties to assist the needy may be appropriately lodged in and fulfilled by governmental institutions and guaranteed as political rights. As Boyle observes, "[t]he duties that exist under modern, political modern welfare rights ... are not simply those of individuals but of the polity as a community, and they are carried out by taxation, which is not discretionary ... [but] ordinarily morally obligatory and enforced by public authority" (Boyle, 2001, 216).

Unlike more conservative commentators (e.g., Fortin, 1982; MacIntyre, 1981; Weithman, 1993), Boyle supports the recent tradition in its recognition of welfare rights as political entitlements, though not uncritically. As we have seen, he is skeptical that assertions of the links between such guarantees and "human dignity" provide sufficient philosophical justification of their status. He therefore seeks to articulate with greater precision the steps required to move from Thomas's emphasis on individual moral duties of discretionary beneficence to welfare rights claimable by individuals against centralized government institutions. There are several elements to his discussion that warrant further scrutiny.

The core emphasis in Boyle's defense of welfare rights remains its grounding in natural law. In his words, "The requirement of common use—that things are to be used to serve human utility in a fair way—is a very basic principle of natural law" (Boyle, 2001, 218). Thus,

> The tensions between political welfare rights and property as justified by Aquinas's arguments are between welfare rights and some values (such as owners' discretion and effective use of things) secured by ownership conventions. They are not tensions between welfare rights and common use as such, for political welfare rights are a way of securing one aspect of common use—sharing resources with the needy (Boyle, 2001, 216).

Still, because Boyle agrees with Thomas's conclusion that private ownership "ordinarily serves common use,"

> ... if the establishment of welfare rights undercuts the effectiveness that ownership provides in securing human benefit from things, causes disharmony among people, or deprives people of the discretion needed for living a creative and morally good life, then establishing these rights will be morally questionable (Boyle, 2001, 216).

As a result, Boyle develops a moderate position concerning the three issues he identifies in the quote immediately above. First, he cautions that a program of welfare rights "should not seek to replace ownership as the basic way humans make good use of things" (Boyle, 2001, 216). This caution, of course, is very much in keeping with the tradition of Catholic social teaching, which both justifies and constrains private holdings in light of the common good. Second, Boyle argues that a system of welfare rights need not provoke social disharmony in the way that Aquinas's own discussion may have suggested, because such entitlements are not conceptualized as common holdings. Instead, as Boyle says,

> The specific source of quarrels that Aquinas thought common holdings caused does not exist, for under systems of welfare rights, goods are not treated as common but as the property of owners or of the needy to whom they are provided as a matter of right. Welfare rights are not a form of access to a commons—rather, they are grounds of duties, held by those controlling resources, to use those resources fairly to benefit the needy (Boyle, 2001, 217).

Finally, Boyle concludes that a system of welfare rights "is compatible with providing a level of discretion over their lives that is needed for developing a virtuous and creative life" (Boyle, 2001, 217). On the one hand, echoing the concerns of mid-twentieth century Catholic social teaching, he acknowledges significant limits on the provision of welfare guarantees. Collectivist systems that overextend welfare rights at the expense of individual discretion threaten the order of charity central to Thomas's thought and to the later tradition. On the other hand, Boyle finds arguments that posit some necessary tradeoff between the exercise of private virtue and the public provision of benefits to be unpersuasive. As a result, he offers several judgments. In general, he suggests, "the experience of modern politics indicates that it is possible for societies to tax at levels that provide significant help to the needy without removing from taxpayers' lives the discretion that they need for virtuous living" (Boyle, 2001, 218). More specifically, Boyle moves to an expansive or *positive* implication of subsidiarity often overlooked by those who view the principle in limiting or negative terms. According to Boyle, the instantiation of a "social" duty of providing welfare as a political entitlement is justified by the necessary role of political action in meeting the requirements of the common good in developed societies. As he says,

> In the specific case of welfare rights, political society uses its specific legal authority to set up social conditions, primarily regulations and redistributive tax laws, in which owners are facilitated in carrying out a basic social responsibility that could not otherwise be carried out so effectively ... The social responsibility of owners ... is rooted in their having a surplus and in the welfare of the needy ... By requiring owners to pay taxes, political society enforces this right of the needy against owners who might otherwise refuse to act responsibly; in this way, political society vindicates the rights of the needy and gives their entitlements legal form (Boyle, 2001, 224).

11.4 Public Provision and Private Virtue

Some commentators (e.g., Weithman, 1993) argue forcefully that Thomas provides no basis for such higher levels of redistributive authority, and that the moral duty to assist the needy should remain an obligation incumbent on individuals or voluntary associations, in keeping with the principle of subsidiarity. On this reading, one finds little or basis in Aquinas's discussion for any later instantiations of welfare claims as political rights, although their status as moral rights remains meaningful. Subsidiarity, as we have seen, implies that the duty to assist the needy is best fulfilled at the lowest level appropriate to the effective implementation of such obligations. Yet this functional aspect of subsidiarity is not its most salient feature. Rather, for such critics, as for Boyle, subsidiarity substantively reflects the appropriate ordering of the life of virtue: human flourishing is better served by forms of association that involve active and interpersonal cooperation.

Nevertheless, as a practical matter, the notion that positive entitlements, instantiated as political rights, would thereby foreclose continuing opportunities for the exercise of individual virtue seems overdrawn.[3] As we have seen in Aquinas's discussion, especially in his comments on property in "On Kingship," Thomas allows the ruler to redistribute resources in enforcing the moral obligation to assist those in dire need. Thus, the differences between Boyle's argument for welfare rights as political and those of commentators who find such an extension of Thomas's perspective illicit emerge from their different empirical judgments about the range of opportunities still available for individuals to exercise personal virtues of generosity and charity amidst such centralized redistribution. After all, any defense of welfare rights that looks to Aquinas's discussion will emphasize the common use requirements of property at stake in meeting the basic or subsistence needs of others. In Boyle's judgment (and mine; Lustig 1993), more than ample opportunities will doubtless remain for the exercise of corporal and spiritual works of mercy, even with a basic welfare system in place.

To say, then, as does one commentator, that discretionary opportunities to help the poor "are not present in a welfare state with a fixed tax rate used to fund entitlement programs" (Weithman, 1993, 174) appears, in light of Boyle's defense of welfare rights, a curious and unconvincing zero-sum version of virtue theory. I should also note, as a final word, that it is emerges as a judgment peculiarly at odds with the thrust of Catholic social teaching. The same Leo XIII, who is, on the one hand, a strong defender of private ownership, is an equally strong proponent of governmental involvement to safeguard the interests of those disadvantaged by modern socioeconomic circumstances. Thus Leo finds no contradiction in supporting both private property and welfare entitlements as political rights. As the founding father of the modern social encyclical tradition, he voices the general sentiment of the popes who follow him throughout the subsequent century when he says:

> The richer classes have many ways of shielding themselves, and stand in less need of help from the State; whereas the mass of the poor have no resources of their own to fall back upon, and must *chiefly depend* upon the assistance of the State [my emphasis]. And it is for this reason that wage-earners, since they mostly belong to the class of the needy, should be especially cared for and protected by the government (*Rerum Novarum*, no. 29).

Notes

1. Any discussion of a moral agent's duty to assist needy others involves, *ab initio*, the consideration of how individual ownership authority is justified. Why? Because any duty I may have to share my resources with a needy person presumes that I, as a moral agent, am justified in my initial appropriation of such resources. I cannot have a duty to meet the needs of others if I am not materially, morally, or juridically in a position to do so. As a general maxim, ought implies can, never more so than in the instance of the justification of private property.
2. As Boyle acknowledges, his own discussion amplifies Thomas's own: "If discretion with respect to property is a necessary means to or an element of the human good, then that is another reason, besides those presented by Aquinas, why natural law grounds the judgment that a person should own some things. I will assume ... that the natural law account of property includes the idea that ownership not only facilitates the proper use of things and promotes harmony, but also provides a form of discretion that is a component of many dimensions of a good life, including virtuous living itself. I believe that this element of the

natural law account is important in thinking about welfare rights and their limits" (Boyle, 2001, 214).

3. One *reductio* of that judgment seems readily available in Thomas's discussion. Think of what Thomas says about those who live the counsels of perfection. As he notes in II-II, Q. 186, a. 3, the vowed religious, in renouncing private property, show *greater* virtue than those who give alms from their abundance. Vowed religious (at least in principle) hold their property in common and exercise a common dominion according to need. And although he deems it unlikely, Thomas in principle leaves open the possibility that all could live this way, with God's grace; that is, as we have seen, Thomas sees no globally *necessary* connection between the Fall and private holdings. Rather, his arguments emerge largely at the level of practical accommodation to individual psychology and societal harmony rather than as fundamental moral or theological claims. For those who claim a *necessary* connection between individual discretion in property holding and the exercise of virtue, what should we conclude concerning the present status of those living the counsels of perfection? Given their renunciation of private property, are they now foreclosed from the exercise of some measure of virtue available to their more ordinary fellows? Is their charity lessened in its scope because they now lack the individual abundance required to exercise virtue toward the needy? One would think, instead, that individual living the counsels of perfection remain more, not less, virtuous on Thomas's account, and not simply as the result of their initial renunciation of worldly goods. Rather, their ongoing disposition to attain the perfection of charity reveals their superior state. Hence, their exercise of virtue toward the needy, though not a material sharing, may prove to be of a different and better variety.

References

Aquinas, T. *Summa Theologiae (ST)*.

Boyle, J. 1996. Catholic social justice and health care entitlement packages. *Christian Bioethics* 2:284–286.

Boyle, J. 1994. Duties to others in roman catholic thought. In *Duties to others* eds. C. Campbell and A. Lustig, 73–89. Dordrecht: Kluwer.

Boyle, J. 2001. Fairness in holdings: A natural law account of property and welfare rights. In *Natural law and modern moral philosophy* eds. E. Paul, F. Miller, Jr., and J. Paul, 206–226. New York, NY: Cambridge University Press.

Finnis, J. 1998. *Aquinas: moral, political, and legal theory*. Oxford: Oxford University Press.

Finnis, J. 1980. *Natural law and natural rights*. Oxford: Clarendon.

Fortin, E. 1982. The new rights theory and natural law. *Review of Politics* 44:590–612.

Lustig, A. 1991. Natural law, property, and justice. *The Journal of Religious Ethics* 19: 119–149.

Lustig, A. 1993. Property, justice, and the common good. *The Journal of Religious Ethics* 21: 181–187.

MacIntyre, A. 1981. *After virtue: A study in moral theory*. Notre Dame: University of Notre Dame Press.

McDonald, W. 1939. *The social value of property according to St. Thomas Aquinas*. Washington, DC: Catholic University Press.

Parel, A. 1979. Aquinas's theory of property. In *Theories of Property: Aristotle to the Present* 88–111, eds. A. Parel and T. Flanagan. Waterloo, ON: Wilfrid Laurier Press.

Schlatter, R. 1951. *Private property: The history of an idea.* London: Allen and Urwin.

Tuck, R. 1979. *Natural rights theories: Their origin and development.* Cambridge: Cambridge University Press.

Tully, J. 1979. The framework of natural rights in Locke's analysis of property: A contextual reconstruction. In *Theories of property: Aristotle to the present* eds. A. Parel and T. Flanagan, 115–138. Waterloo, ON: Wilfrid Laurier Press.

Weithman, P. 1993. Natural law, property, and redistribution. *The Journal of Religious Ethics* 21:165–180.

Chapter 12
Health Care Technology and Justice

Germain Grisez

12.1 The Need for Health Care

The title of this paper refers to a subject matter that is already being dealt with by many able scholars, for some of whom it is a central preoccupation. The issue of justice in health care has also been an abiding concern of Joseph Boyle (see, e.g., Boyle, 1977, 1996, 2001). In essays ranging over some 30 years, he has defended a right to health care in developed nations, and has addressed some of the difficulties that arise in a context of finite resources and moral pluralism. In this essay I shall only propose some ideas that I hope will be helpful to Boyle and others who wish to contribute to the ongoing debate on these matters.

Health care here does not refer to contraception, abortion, in vitro fertilization, freezing embryos and corpses, sex change surgery, cosmetic surgery, assisted suicide, euthanasia, or drugs or procedures meant to enhance an already-healthful somatic or psychic function or to reduce the discomfort of healthful somatic and psychic states, such as fatigue and grief.

In health care, the physician-patient and nurse-patient relationships are central, but, in addition to physicians and nurses, many others help meet health care needs: dentists, optometrists, pharmacists, hospitals, community clinics, pharmaceutical companies, phlebotomists, respiratory therapists, x-ray technicians, critical care paramedics, nurse practitioners, and so on and so forth. When referring in general to those who help meet health care needs, I call them "providers," and I call those whose needs they meet "clients." In using these general expressions, I do not mean to reduce providers and clients to their roles in the relationship of supplying and obtaining things that cost money, nor do I mean to disparage the dignity of any kind of provider.

The justice of existing laws and present practices regarding many matters, including taxation and immigration, is questionable and vigorously debated, and reforms in any of these fields would affect all the others. But in what follows I focus on

G. Grisez (✉)
Mount St. Mary's University, Emmitsburg, MD, USA
e-mail: grisez@msmary.edu

C. Tollefsen (ed.), *Bioethics with Liberty and Justice*, Philosophy and Medicine 110,
DOI 10.1007/978-90-481-9791-0_12, © Springer Science+Business Media B.V. 2011

justice in health care and prescind from other matters. So, my statements about what justice requires for citizens and lawful residents should not be read as saying anything about what it requires for illegal immigrants, and my remarks about taxes should not be read as endorsing existing tax laws and their enforcement or any proposal to change them.

All knowledge about how to care for one's own and others' health can be called *health care technology* in a broad sense. In this sense, the people of every society acquire by experience and hand on a health care technology—a body of knowledge about how to identify what is safe to eat and drink, how to protect themselves against environmental challenges, how to help birthing mothers, how to care for babies and the incapacitated, how to avoid getting and spreading diseases, how to avoid injuries, how to deal with illnesses and injuries, and how to function despite them, and despite handicaps and declining capacities. Without such practical knowledge, people could not survive. Their common moral responsibilities with regard to this technology are to seek that practical knowledge, share it with others, and cooperate in acting on it.

During modern times, health care technology greatly increased wherever the means of communication improved, formal education became widespread, and systematic inquiry was conducted in biology, chemistry, and health-related statistics. As people in industrialized societies also grew wealthier, they became more able to act on their increasing knowledge—for example, by providing clean water, sewage disposal, a better balanced and more adequate diet, and more adequate clothing and shelter for themselves and their families. As people's ability to live healthily increased, infant mortality declined, the spread of communicable diseases slowed, and life expectancy rose.

In most if not all societies, certain people with special responsibilities for providing health care have possessed and handed on specialized technology. With the modern development of health care technology, such people acquired many new tools and behaviors: drugs, devices, and procedures for use in prevention, diagnosis, and treatment of diseases and injuries. With the increasing variety of skills required to use burgeoning technology, health care professions—or jobs—multiplied and were diversified. Businesses were founded to develop, manufacture, and market drugs and devices; and hospitals, which previously offered nursing care to poor people, became places designed and equipped to apply a broad range of the ever-expanding health care technology. Health care thus became a complex industry, and *health care technology* now is generally used in a narrow sense to refer to the set of goods and services that industry provides.

In industrialized nations during the past two centuries, technological innovation in many fields has been increasingly rapid, and there also has been rapid innovation in the fine arts, literature, and various forms of play. Because of human creativity and the unquenchable human desire for new goods and services, no limit to such technological and cultural innovation is or even can be in prospect. Nevertheless, most natural desires could be satisfied adequately without further innovation. For example, most people could have an adequate diet without taking advantage of innovation in processing and marketing foods and drinks, and in preparing and presenting meals. Moreover, frugal people generally delay or even entirely forgo taking advantage of innovations that themselves generate desires for new products and services rather than respond to some antecedently perceived need.

In contrast, with such adventitious desires and even with other natural desires, the desire to preserve life and bodily integrity, and to maintain and promote healthful functioning, generates demands for innovative technology. For while it can be easily satisfied by most people during part of their lives, almost everyone sooner or later urgently needs up-to-date products and/or services of the health care industry; and many people would benefit even more from anticipated technological innovations.

Sometimes health care technology decisively wins a battle, as when systematic vaccination eradicated small pox. But many elements of health care technology have unwanted side effects and/or limited effectiveness. Then improvements are desired. Moreover, as life expectancy has increased, additional, treatable physical and psychological pathologies have been identified. Currently, the study of genes, molecular biology, and cell biology promise breakthroughs in the diagnosis and treatment of many diseases, and in the growing of replacement tissues and organs. If the human life span is naturally limited, as some maintain, to about 120 years, present health care technology still has a long way to go before that limit is approached by the life expectancy of people whose lives are not shortened by violence or injuries. Moreover, there seems to be no end to the possibilities for dealing more effectively with the damage to bodily integrity and functioning caused by injuries. Therefore, although, as in other fields, no end is or can be in prospect to innovation in health care technology, innovation in this field is peculiar in being driven by an urgent natural desire that at times in most people's lives cannot be fully satisfied.

Due to ever-expanding technology and the unpredictability of the need for health care, that need, unlike others that are clearer and more definite, is indeterminate and always growing. This peculiarity of health care technology significantly contributes to the difficulty of questions about moral responsibilities in its regard. Although there are other questions, I will focus on those about providing and paying for health care.

With various exceptions, including many drugs, people have never been able to shop for health care technology as they do for most other goods and services. Very often, people with incipient health problems do not know what care they need. Diagnoses often are uncertain, and alternative plans of treatment often are incommensurable with respect to their prospective benefits and burdens, including their prospects of success and risks of bad side effects. Providers often must motivate the demand for the goods and/or services they will supply. Clients often are more or less committed to a provider before he or she makes a diagnosis or proposes a plan of treatment, which usually is tentative, sometimes must be worked out step by step, and seldom is sure to succeed. Reliable information that would be needed to compare potential providers is often unavailable, and nonmonetary factors often impede or prevent clients from changing providers. Consequently, providers seldom compete by offering at different prices health care of similar quality to meet the same need in the same way. Thus, there never has been and never will be a market for health care technology comparable to the markets for groceries, automobile servicing, and hotel rooms. In fact, many providers have no effective competitors, because there is no alternative provider reasonably available to their clients. Realizing how matters stand, providers are likely to be tempted to set exorbitant prices for their goods and services.

Until the twentieth century, however, physicians and nurses mainly offered advice, moral support, nursing care, and the alleviation of symptoms. Most people seldom urgently needed such professional care, and many went through life without ever receiving it. When a form of care that would affect survival or functioning was clearly needed and urgently desired, its cost was seldom burdensome for the wealthy and influential. Some people who could not afford clearly needed care obtained it from charitable institutions or providers who waived part or all of their fees, because of their professional commitment to serve and their acquaintance with community members who needed their service. Of course, some poor people died due to lack of professional health care, but their deaths were regarded as part of the normal course of events just as were others' deaths due to lack of an adequate diet.

As the health care industry took shape during the twentieth century, technology regarded as effective by most clients proliferated, and people with health problems increasingly desired that technology. At the same time, health care grew more and more expensive, causing many clients to fear they would be unable to pay for it when they or their dependents clearly needed it.

Never able to care for all poor people, charitable institutions could not begin to meet the growing needs of an even larger part of society. As the cost of other elements of care grew, the willingness of a provider to reduce or waive fees fell increasingly short of solving the problem of a client who could not afford needed care. Moreover, some providers no longer had the professional commitment common in earlier times, and increasing mobility together with the differentiation of health care jobs often meant that providers had no direct relationship at all, or only a transient one, with their clients.

12.2 Schemes for Meeting the Need

As a result of these complex developments, by the mid-1930s many people in industrialized nations were anxious about how to meet their own and their dependents' needs for health care. By then, too, there was a large middle class with social and political influence in all these countries. Many people were therefore ready to welcome and able to promote arrangements to avoid incurring fees for health care at the time it was needed—fees that might well be beyond their means, exorbitant, or both. At the same time, many providers were open to ways of ensuring timely payment for their products or services, without limiting their clientele to those who could pay for what they needed when they needed it. Thus, beginning in the 1930s and continuing through subsequent decades, many schemes developed for providing and paying for health care.

All those schemes can be understood as variations on two basic models: insurance and prepayment. In the former, for the payment of a premium, the insurer agrees to pay or reimburse providers for a more or less extensive set of health care products and/or services when they are needed. The latter takes two forms. In one, a provider or set of providers of health care products and/or services undertake to meet, more or less, the needs of a specified person or set of persons. In the other, an organization undertakes to meet the needs of a specified set of persons by employing providers and providing facilities.

With either basic model, the payer can be the covered individual (or family), a government for all or some of its citizens, another entity (such as an employer), or some combination of these.

Until Section 12.5 below, I will not deal with insurance schemes that undertake only to pay or reimburse for certain sets of costly products and services that most covered persons will never need, nor will I deal with prepayment schemes that guarantee only to deliver specified products or services at specified times. I will deal with those that undertake, subject to various conditions, to provide or pay for a broad variety and unspecified quantity of forms of care.

Most schemes have involved some sort of mixture of the two basic models. For example, an insurance scheme may specify certain products and services, such as inoculations and checkups, to be supplied to any covered person who wants them, and a prepayment scheme may promise reimbursement within limits for products or services obtained from third parties.

Most schemes limit and some designate the providers who will be available, and all employ managers to ensure that providers and clients meet conditions set by the scheme for delivering what it promises. No scheme undertakes to meet every clear need with a free choice among all the products and services that health care technology as a whole makes available.

Regardless of differences among schemes, the possibility of obtaining care without incurring any—or any additional—expense makes it easier to seek that care. Consequently, clients will count on obtaining care they otherwise would not expect, and clients are likely to obtain care they otherwise would not obtain. Clients thus have a sense of entitlement to care.

In some cases, that effect on clients' motivation is all to the good, for example, when it leads them to get appropriate inoculations and check-ups they otherwise would neglect. In other cases, it is all to the bad, for example, when it leads nervous people to obtain risky care they do not need or leads people who otherwise would maintain a healthful lifestyle to overeat, use tobacco, abuse alcohol, carelessly risk transmitting or contracting diseases, and/or neglect hygienic practices, such as washing hands.

Probably far more often, the possibility of obtaining care without incurring any (additional) cost leads people with a reason to desire some form of care to obtain it even though their need for it is not clear and, if they had to pay directly for it, they would delay seeking it or prefer a less expensive alternative. For example, if new drugs are covered, many clients will prefer them to generics; if heart scans are covered, many healthy clients will prefer them to stress tests; if MRI examinations and spinal surgery are covered, many clients who would otherwise deal conservatively with back pain will prefer an MRI for diagnosing a ruptured disk and surgery for treating it.

More generally, coverage by any scheme will motivate many clients to see their primary care physician or go to an emergency room before symptoms are intense or definite enough to make clear the need for care; and a primary care or emergency room physician's diagnosis of the onset of a chronic condition will more quickly lead many clients to a specialist for a complete work up, a plan of treatment, and regular return visits. Thus, more clients with sore throats are quickly checked for streptococcus despite the lack of other symptoms; more clients receive eye examinations by ophthalmologists rather than optometrists; and more clients with digestive problems soon see gastroenterologists, receive complete examinations, and return for regular check ups. Such accelerated choices to seek and extend treatment occasionally prevent disasters and seldom lead to serious regrets.

Since choosing to obtain care without incurring any, or any additional, cost seems reasonable and is not obviously harmful to others, most people who make such choices will not ask themselves whether that is fair to whoever must ultimately pay for this care that they would not obtain if they had to pay for it themselves. Conscientious people who do ask themselves that question reasonably judge that, since others covered by the scheme take full advantage of what it offers, they are not duty bound to deny themselves any available benefit as long as they proceed honestly within the scheme's rules.

Clients' sense of entitlement motivates most individual providers to supply more products and/or services than they otherwise would, partly because most want to be accommodating and partly because they do not want to be blamed for the occasional bad result of not supplying what a client wanted. Moreover, under any scheme, some providers are tempted to supply or to say they are supplying more goods and/or services than clients desire or need, in order either to meet performance standards set by managers of prepayment schemes or to increase their income in insurance schemes.

Consequently, no matter how a scheme is structured, its costs tend to exceed the amount that those ultimately paying are prepared to pay, and the scheme's designers and managers must limit what and how much it supplies or pays for. Since many of the items delivered or paid for are individually small in cost, managers cannot bring about the needed limitations by closely monitoring each and every transaction, but they can and do try in other ways to put the lid on.

12.3 The Shortcomings of Existing Schemes

Every sort of scheme can try to limit how much it will provide by requiring a manager's or management-designated agent's prior approval (or preauthorization) for the use of certain elements of health care technology. That requirement promotes the common good and harms no one insofar as it reduces waste due to some clients' unreasonable desires and/or some providers' excessive deference, bad judgment, or greed. However, when people without first-hand knowledge of clients and their needs are responsible for setting limits, they are likely to be tempted to do that without adequate consideration and due regard for clients' best interests. A client's failure to obtain required prior approval, due to forgetfulness or temporary non-competence, also can result in a serious loss. At the same time, providers and clients are likely to be tempted to exaggerate—that is, to lie about—the seriousness and urgency of needs so as to increase the likelihood of obtaining approval or authorization for care they consider necessary.

By promises and/or threats—usually implicit rather than clearly articulated—a scheme's managers also can try to motivate providers to choose to omit some examinations and treatments, to refuse to write some prescriptions, and/or to avoid making some referrals. This also promotes the common good and harms no one insofar as it reduces waste. But even without a provider's being aware of it, this pressure from managers can color his or her views of what is in clients' best interests and thus lead to judgments at odds with them. And insofar as promises and/or threats lead competent providers to make choices contrary to their judgment about what is in a client's true interests, they are led to betray their professional responsibility and deprive the client of the loyal service to which he or she is entitled. Clients who are less aggressive and less able to look after their own interests are of course more likely to suffer in this way.

How much a scheme provides can also be limited by charging copayments—specified amounts that clients must pay towards various products or services. Insurance schemes also can specify both deductibles (amounts that must be paid before coverage or some elements of it begin) and limits on the total coverage or some of its elements. Just to the extent that clients desiring care encounter such limits, the scheme does not benefit them; yet participation in the scheme remains more or less constraining to them as well as to potential providers.

The common interest in keeping health care affordable may seem to require copayments, deductibles, and limits of coverage. However, the more someone can afford to spend, the less effective these disincentives are; so, the less likely they are to prevent choices at odds with the common good. The less someone can afford, however, the greater are the disincentives of such costs and limits to obtaining even clearly needed care; and the more likely they are to lead to choices at odds with the best interests of those making them and/or of their dependents. So, such methods of limiting what a scheme undertakes are unfair, because they mainly limit what poorer participants get out of it, and may well deter them from obtaining care they clearly need.

Managers of a scheme involving the insurance model can limit payments in several ways besides those already mentioned. Long and complicated contracts not made available to individual policyholders can make it virtually impossible for them to know what they are entitled to. The process of filing claims can be made complex and inconvenient for providers, those covered, or both. Managers can use technicalities to generate plausible excuses for refusing, reducing, or delaying payments; they can require precise and even repeated submission of evidence to support claims. Subordinates can be given subtle incentives to misplace or lose paperwork. Due to the use of such tactics, some legitimate claims are never filed, some that are filed never meet technical requirements for approval, and some that meet the requirements are abandoned in frustration and despair of ever being paid. At the same time, such tactics waste policyholders' and providers' time and require the latter to hire people or pay agencies to file insurance claims and handle problems with them.

Managers of a scheme involving the prepayment model directly limit the goods and services it makes available by supplying what their limited budget can buy while managing access so that participants receive a more or less fair share of the limited quantity of products and services. The more necessary it is to restrict access to the promised benefits, the more likely participants are to experience the unavailability of some elements of health care technology, delays in receiving goods and services, overcrowding in facilities, and other inconveniences and discomforts.

Managing access necessarily involves scheduling care on the basis of a judgment about whose needs and/or which needs deserve priority. While it may be said that the judgment is made by considering which need is more serious, more urgent, and more likely to be met effectively by prompt care, it is likely to be influenced by other factors. Other things being equal, for instance, participants whose prospects for recovery make their lives seem more worth saving or whose needs can be met quickly and easily may be given priority. In any case, the result of such scheduling is that those without priority will receive care later rather than sooner, if they receive it at all. For example, cataract or hip-replacement surgery might be forthcoming only after a long wait.

Managing access also means limiting care for chronic conditions and accepting some risks of failing to adjust treatment when appropriate or to respond in timely fashion to new symptoms calling for a radical change in the plan of care.

Finally, managing access can mean refusing to provide a means of care despite its foreseen benefits if the benefits are judged inadequate to justify the burdens, including the impact the care's cost will have on the scheme. As clients' life expectancy and chances for recovery lessen, the prospective benefits of providing care for them decrease and its costs often increase. Those managing access to care may thus decide not to supply costly examinations and treatments to participants who are old, gravely debilitated, or both. Such participants may receive only less costly treatments and medications, perhaps only palliative care.

Because delays and limits increase participants' risks and suffering, they—and any allies they may have among providers—are likely to be tempted to conceal relevant information and to lie in order to obtain more care or obtain care sooner than competing claimants. At the same time, because judgments prioritizing needs involve many incommensurable variables, the rationality of such judgments often is unclear, and those making them may well be unfair. The unfairness might be due to some factor which is not even consciously considered (for example, personal experience with similar problems), or due to rationalization (for example, admiration for a client's courage and cheerfulness). Or it may be a deliberate and admitted unfairness—for example, due to a personal relationship, or a bribe or threat.

The managers of some schemes structured on the prepayment model identify the minimum payment sufficient to obtain from some potential providers each product and service the scheme promises to those covered. The managers then offer payments only slightly above that minimum, but do not obtain commitments from or otherwise motivate any group of potential providers to supply or serve all those covered at the specified prices. When those covered can find no provider ready to supply what was promised at the payment offered, this tactic unfairly breaks the promise. It also may tempt providers to bill for more than they provide, accept side payments (bribes) from those needing care, and/or reduce the quantity and/or quality of what they provide.

Despite the proliferation of schemes, most clients obtain at least some of their health care by paying directly for it. People not involved in any scheme must do so or do without the health care they need, unless they can obtain it without charge. Because no scheme undertakes to meet every clear need with a free choice among all the products and services that health care technology as a whole makes available, most clients covered by a scheme also must pay directly for some health care. The wealthy sometimes pay for what they regard as better and/or more convenient care, and even those who are not wealthy sometimes do this in order to obtain some urgently desired element of care more quickly and/or from a preferred provider.

12.4 Injustices Peculiar to the United States

In the United States, many people are not covered by any health care scheme—some because no scheme will cover them, some because they cannot afford coverage, some because they opt out of an affordable private scheme or fail to apply to a governmental scheme for which they are eligible. The schemes are quite diverse, and most do not remain unchanged for long. Despite the efforts of designers and managers to limit what and how much schemes provide or pay for, the ultimate payers are constantly pressed to pay more. Yet because of efforts to limit costs, most providers as well as many clients consider themselves to be involved in an arrangement unsatisfactory not only in what it does for them but also, and even more, in what it imposes on them and requires of them.

By allowing employers and self-employed individuals to pay with pretax dollars for coverage by health care schemes and to deposit pretax dollars in health care reimbursement accounts, the U.S. government motivates many people to maintain coverage by such schemes and/or to set up such accounts, and thus subsidizes the health care of all who consequently pay less tax than they otherwise would. This subsidy is greater for those whose income is higher, because they pay taxes at a higher marginal rate and generally have more costly health care schemes and/or larger reimbursement accounts. Covered people whose income is so low that they owe no tax on it do not receive any subsidy on this basis, nor do employees whose employers neither provide a health care scheme nor offer health care reimbursement accounts.

Currently (2008), U.S. citizens and permanent residents become eligible for health care coverage by Medicare at age 65 if they or their spouses have had at least forty quarters of employment covered by Social Security. Most do not pay for Medicare part A (hospital insurance). But for Medicare part B (medical insurance) there is a charge, ranging from \$192.80 per month (for a married couple whose household income is under $164,000 per year) to $476.80 per month (for a couple whose household income exceeds $410,000 per year). This scheme also covers certain other people under various conditions—a couple without any covered employment can obtain Medicare parts A and B for $1,038.80 per month.

Besides the payment required to obtain it, Medicare has many gaps, which participants can more or less fill by buying one or another of a standard set of medigap policies from a commercial insurance company. So, Medicare is far less adequate for those who cannot afford a medigap policy than for those who can, most of whom also spend a smaller percentage of their greater incomes on the basic Medicare coverage.

In the United States, citizens and permanent residents who are very poor, regardless of age, are eligible for coverage by Medicaid. Thus, many very poor elderly people in the U.S. are eligible for coverage by both Medicare and Medicaid. However, some people who are ineligible for Medicaid—because, though far from wealthy, their earnings or assets exceed the eligibility limits—have no other coverage, and can afford neither to purchase basic coverage (or a medigap policy, if covered by Medicare) nor to pay directly for health care they clearly need. Consequently, in the U.S., some people who work hard but earn little never in their lives receive governmental help to obtain health care they clearly need but cannot afford, while most people who are better off receive at least some help during part of their lives. Those who generally receive the most governmental help during their lives as a whole are those who need it least: people who have high earnings during their working years and substantial household income in retirement. (Although the small proportion of couples whose household retirement income exceeds $164,000 per year must pay more than the minimum for Medicare, that payment still is far less than the cost to the government of their Medicare benefits, and their higher payment for Medicare often is more than offset by the greater tax savings they enjoyed on nontaxable coverage during their working years.)

Medicaid, funded partly by the federal government and partly but variably by each state, generally pays less than Medicare or private coverage schemes pay for similar services. One consequence is that fewer physicians accept Medicaid, and those covered by it sometimes cannot find individual providers willing to care for them. However, federal law requires any hospital that accepts Medicare and operates an emergency room to examine anyone who comes to that facility and to care for such people until their condition is stabilized. So, poor people who have Medicaid coverage and people without any sort of health care coverage often seek care from hospital emergency rooms. Since there is no governmental payment for people without coverage and since Medicaid payments generally are less than the marginal costs of emergency room service, most hospitals incur substantial losses on their emergency rooms, which often are overcrowded with people who must wait many hours for evaluation and care.

Hospitals compelled by the law to provide emergency room care without charge, as well as other providers of health care products and services covered by Medicaid and Medicare, usually set or negotiate higher prices for the same things when providing them to someone covered by a nongovernmental scheme. But providers often set the highest prices for clients who, having no coverage, incur the obligation to pay for anything they or their dependents get as they get it.

Such cost shifting especially burdens the latter group. The burden falls on them not because they are well suited to bear it but simply because imposing it on them is possible and their need compels them to pay what is demanded. The burdened group includes the working poor who have no coverage, and who consequently sometimes forgo clearly needed care they would get if they were either wealthier or less honest and industrious. Insofar as the burden of cost shifting is deliberately imposed by governmental policies and actions, it is morally equivalent to an unjust tax; insofar as it is deliberately imposed by hospitals and other providers, it is morally equivalent to theft.

12.5 Towards a Just Health Care Scheme

Since all existing health care schemes are not only unsatisfactory in other respects but morally problematic, and since much of what the U.S. government does about health care is blatantly unjust, U.S. citizens and governmental officials share a grave moral responsibility to seek and promote justice in providing and paying for health care.

I doubt that there is a single, just resolution to this problem. There are likely to be several morally acceptable resolutions, each with incommensurable advantages and disadvantages of its own. Moreover, only groups including experts in health care, economics, law, and politics are likely to be able to work out practicable resolutions to the problem. But some requirements for establishing justice in health care can be formulated, and I hope my attempted formulations may help guide efforts to develop morally acceptable and practicable resolutions.

First, governments must be involved not only in setting and enforcing standards for health care, and regulating its provision and payment for it, but in paying for health care for certain groups of people, including those engaged in public service whose health is essential for their service (for example, military personnel) and those whose well being as a whole is the responsibility of public authorities, such as prisoners and incompetent persons for whom no one else is responsible.

The proliferation of health care technology and the social and economic structure of the health care industry make it no longer reasonable to expect providers regularly to reduce or waive their charges for care supplied to poor people. Insofar as providers are compelled to provide care without charge, they are induced to shift costs indiscriminately, and therefore unfairly, to those who can pay for care. The health care needs of poor people also far outstrip the resources available to non-governmental institutions, such as churches and charitable organizations. Therefore, governments ought to see to it that hospitals and individual providers who care for poor citizens and lawful residents receive fair compensation.

How much and what sort of health care of poor people should governments pay for? Other things being equal, a government capable of it surely ought to spend at least as much on the health care of poor people as on the health care of people for whose care as a whole it is responsible. But how much governments should pay and for what can only be discerned by considering available resources and competing needs, and applying the Golden Rule. In doing so, two things should be borne in mind: first, that the need of poor people for health care is not less real or important than that of people who are wealthy; second, that meeting that need and certain other needs of poor people, such as their need for education, would enable many to escape from poverty and use their gifts to contribute to the common good.

Second, governments must set the prices they will pay health care providers for the goods and services they deliver to poor people, who at times need almost all the products and services provided by the health care industry. Governments already set prices for most—for example, the U.S. government sets prices for everything covered by Medicare. Of course, the prices may not be fair, and setting fair prices would be a complex and costly undertaking. However, to fulfill their responsibility to citizens and to avoid injustice, governments must set fair prices for every product and service provided by the health care industry. For, if governments pay too much for those products and services, the common good is injured; if they pay too little for any of them, providers and/or those to whom providers shift costs are cheated.

Often the fair price will be more than the minimum most providers would accept for caring for poor people. Most providers of health care products and services will at times deliver them for a payment covering slightly more than their marginal cost, since that seems profitable inasmuch as it is assumed that other payers have covered or will cover other costs. However, if a government in setting prices on the health care products and services supplied to poor people takes advantage of that tendency of providers, it unfairly shifts to other payers some of the capital cost of health care facilities, the training of health care personnel, providers' business overhead, and so on.

At the same time, the fair price often will be less than the price a provider would otherwise charge and could obtain. After all, monopolies can and often do charge exorbitant prices for things people urgently need, and, in an economic even if not in a legal sense, some providers have a monopoly on elements of health care technology: those who hold a patent on a drug or device and those who alone can provide a service needed here and now. Moreover, as already explained (in Sect. 12.2 above), many providers have no effective competitors. Fair governmental price setting for elements of health care thus will require a procedure similar to setting the rates charged by public utility monopolies: the full costs of each item must be determined and allocated, and a reasonable margin of profit allowed.

If governments set and pay fair prices for health care products and services, and providers charge other payers more for the very same things, the excess charge will be unnecessary and unfair. So, no provider that sometimes accepts a payment or subsidy from the government for a product or service should be permitted to charge anyone more than the fixed price for that product or service.

Third, considerations already set out (in Sect. 12.3 above) show that there are serious problems with schemes that undertake to cover a broad variety and unspecified quantity of forms of care that many of those covered will be more likely to obtain because they are covered. By requiring that managers attempt to limit what is delivered, such schemes tempt everyone involved, including the managers themselves, to act immorally; and, even so, do not achieve satisfactory limitations. It therefore seems to me that such schemes cannot contribute to a just solution to the problem of providing and paying for health care. If I am right, in the United States justice requires not merely change in health care financing but the termination of Medicare, Medicaid, and virtually all the private schemes currently in operation.

Nevertheless, both insurance and prepayment schemes can and must be parts of a just system.

Insurance can soundly cover any set of elements of health care technology that is quite costly and constitutes care known to be both necessary and effective for a condition that will not become much more likely if it is covered by insurance—for example, the elements of care needed by someone who suffers a stroke, perhaps beginning with an ambulance ride to the emergency room and ending with a final visit to a physical therapist. Some chronic conditions also meet the standard for insurance coverage—for example, those resulting from genetic and other congenital defects—but it cannot be met by chronic conditions that different lifestyle choices would very likely forestall.

Because health care technology develops and is subject to revision, it will be an ongoing task to specify the conditions and sets of elements to be covered, as well as the criteria for a diagnosis and prognosis that must be met for appropriately initiating and carrying on the course of care. Besides being inherently difficult, specifying the objects and conditions of coverage also will be controversial, and the stakes will be high for all parties. So, the task might best be assigned to an entity with some insulation from the political process, similar to that enjoyed by the U.S. Federal Reserve Bank.

Insurance also can cover one set of elements of health care technology that is quite costly and constitutes appropriate health care for a condition likely to become more common if it is covered—the one that typically begins with a pregnancy test and ends with a final postnatal checkup. This coverage is warranted because it subsidizes the care of new members of society and those who give birth to them.

Automobile insurance companies need to be able to tell when they must pay on a claim, make sure that the work they are paying for is done properly, and prevent errors and fraud. So, if soundly managed, they obtain accurate information about accidents and check out damaged vehicles and the work of body shops. For similar reasons, health care insurance requires that those who make payments have access to the care process, all parties to it, and all relevant information. To ensure that the conditions that warrant providing care have been met, preauthorization generally will be required. Those who handle claims will need enough education and experience to deserve the respect of health care providers and to know when to require further evidence and/or second opinions.

Handling claims properly will be time consuming. Those who handle them, being well qualified, will have to be well paid. So, managing the insurance scheme will be costly. Still, the work of good claim handlers will be not only necessary for sound health care insurance but beneficial: it will motivate providers to adhere to sound practices and, in general, be more careful than they otherwise would.

Prepayment can cover a package of elements of health care technology that every covered person will receive because everyone, or every member of one sex or the other, should receive them at certain ages or stages of life, or at certain intervals during at least part of their lives. Most health care providers believe it would be good to do certain things for every person, even those with no symptoms or injuries. Not all providers of a certain kind of service are likely to agree entirely about what all clients need, but surely providers can reach a consensus about some things that will benefit everyone and when those things should be done. Elements can be covered by prepayment if there is both a consensus among providers that they are needed and supporting evidence that they are worthwhile. The package might include various inoculations, some medical screening and testing, and a periodic session with a primary care physician; a periodic dental examination and cleaning; and a periodic eye examination.

Although the worth of annual checkups has been questioned, a periodic consultation with a primary care physician could be cost effective. The interval between consultations might vary at different stages of life. The purpose would be to receive appropriate inoculations, discuss the results of screening and testing, review and record any care received since the last checkup and its results, plan additional care if needed, and deal with matters such as diet, exercise, and the use of nonprescription medications.

Much effective preventive health care consists in supplying information and advice. Effective information and advice must take into account an individual's peculiar condition, needs, and limitations. Well-constructed computerized data bases can supply precise answers to adequately specified queries formatted in a standardized way. Suppose that all the information gathered during the course of someone's receiving health care—including the examinations and tests covered by prepayment, and information about care already received and its results—were kept in an electronic file with a standardized format. That file could be used along with a well-constructed, preventive-care data base to generate personalized advice for presentation by a primary care giver to each client along with appropriate explanations and in as persuasive a manner as possible.

Fourth, everyone clearly needs at least insurance and prepayment coverage of the sorts just described. Poor people cannot pay for that coverage, and nobody except governments can reasonably be expected to pay for it on their behalf. So, if governments can pay at least for that coverage for poor people, they should do so.

If the United States government supplied that insurance and prepayment coverage to poor people, they could obtain more easily much of the care Medicaid now delivers, and that care would be better. (As I shall explain under "Fifth" below, poor people also would need a subsidy to pay directly for care that cannot be covered by insurance and prepayment.) If the U.S. government supplied the same insurance and prepayment coverage, without copayments and deductibles, to the near poor—most of whom now have no coverage at all or only inadequate coverage—most of them would be able to pay directly for other care they need. So would most of the elderly, if the government supplied the same coverage for them, without any copayments and deductibles, in place of Medicare.

If people earning an adequate living had the same insurance and prepayment coverage—again without copayments and deductibles—it would, to a great extent, replace the coverage most of them currently have under various insurance and/or prepayment schemes—coverage mostly paid for by those people and/or their employers with pretax dollars, and in that way, as explained in Sect. 12.4 above, unjustly subsidized by the government. Two questions arise: Who should pay for an insurance and prepayment package, along the lines I propose, for people earning an adequate living? Can the U.S. government justly subsidize it, and, if so, how?

Someone might respond that such people should pay for all of their health care, just as they pay for the food they eat, with post-tax dollars, and that the government should subsidize them with a tax credit inversely proportional to their income. In that case, however, health insurance no longer would be nearly as attractive an employment benefit as it now is. Its only advantage would be that a group of employees can be insured more easily and cheaply than they can as individuals. However, employers either would stop offering health insurance or would offer it as an option that many healthy employees would reject. Affordable insurance would then become unavailable to unhealthy people. Groups without their healthier members would pay dearly for coverage; unhealthy individuals would pay even more dearly or simply be refused coverage.

Some think that such problems could be prevented by governmental mandates. But how can it be just to compel healthy individuals to buy health insurance that they do not think they need? And how can it be just to compel insurance companies to insure those whose claims predictably will exceed what they pay in premiums? Moreover, governmental mandates may well be impossible to enact or be unconstitutional.

Still, except for some very wealthy people, everyone needs health insurance, since most people will eventually need, for themselves or their dependents, care that they will not then be able to afford. Denying it to them would be cruel; compelling providers to supply it without payment would be unjust; and using public resources to supply it would unfairly burden those who had foreseen such needs and prepared to meet them. Thus, the common good requires that people have some fair way of paying for that care. The only fair way anyone has suggested is for the government to make the health coverage of those earning an adequate living a public benefit, and paying for that benefit by increasing everyone's income taxes—a method of payment that also has the advantage of automatically subsidizing those whose income is less. Consequently, it seems to me that the U.S. government and any government that can afford the health care insurance and prepayment coverage that everyone needs should pay from general revenues for the same coverage for all citizens and lawful residents.

In the United States, and perhaps elsewhere, insurance schemes, such as drivers' liability insurance and workers' compensation insurance, have covered health care in certain cases. Most people regard those arrangements as reasonable and just. So,

I do not think governments should accept responsibility to cover the health care required to deal with injuries or illnesses that occur in such cases.

Fifth, insofar as possible, clients ought to pay from their own resources for any health care that cannot be covered by insurance or prepayment and that they or their dependents need. In this way, clients themselves, rather than managers or officials who do not know them well, will judge, with the advice of caregivers, what is in their own interests and will make the hard choices between health care and other goods they need and want. Moreover, the prospect of having to pay for health care from their own resources will motivate most people to avoid many illnesses and injuries.

Poor people must be helped to meet their need for health care as well as their other needs. But poor people either can judge what is in their own interests or not. If they can, they should not be given separate subsidies for food, shelter, and health care. Rather, they should be given a reasonable periodic payment and allowed to decide how to use it. If poor people cannot judge what is in their own interests, appropriate guardians or caregivers should be assigned responsibility for receiving such people's payments and making the necessary judgments on their behalf.

With governmental price fixing, the care for which clients will have to pay will be more affordable than it otherwise would be. Then too, fair governmental payments for the products and services covered by insurance and prepayment will help health care schools and charitable organizations operate community clinics in which fees for other elements of health care will be reduced or waived, and will enable generous individual providers to negotiate appropriate financial arrangements with clients in financial straits. Governments also might subsidize some elements of health care technology—for example, very effective drugs not covered by insurance or prepayment whose fair price is too high for many clients. Some providers could also be subsidized—for example, those serving in poor and thinly populated areas. Governments should never subsidize health care in ways that benefit the wealthy more than poor people.

Governments should pay each provider fairly for health care they legally require that provider to supply to someone who cannot pay.

Sixth, elements of health care technology that are unsafe or ineffective or both are gravely dangerous and often wasteful. The responsibility for judging and certifying safety and effectiveness cannot reasonably be entrusted only to potential providers, and clients or private organizations seldom have information and resources to do the job. Therefore, governments that can evaluate the safety and effectiveness of health care technology should not allow anything to become or remain generally available unless shown beyond reasonable doubt to be safe and effective, and if they directly or indirectly pay for or subsidize health care, they should allow in that care the use of nothing whose safety and effectiveness have not similarly been shown.

Before new drugs, biologics, and devices are approved to be made generally available in the United States, the U.S. Food and Drug Administration (FDA) requires a review process to determine that they are safe and effective when used as directed. Here, *safety* does not mean no bad side effects, but that such side effects can be reasonably accepted or recognized in time to discontinue use and avoid unacceptable results; *effective* does not mean using the item will always have the hoped for good effect, but that using it is likely to have the effect in a specified percentage of cases—or, in the case of certain items approved for use because there is no better alternative, that there is sufficient reason to hope for benefit to make it reasonable to use them.

The FDA's performance in fulfilling its responsibility has been criticized and surely could be improved. But more relevant here is that the FDA's authority is limited: some old drugs and devices were never approved; herbal medications and dietary supplements are not subject to review; many approved drugs are prescribed for off-label uses in which their safety, effectiveness, or both are reasonably doubtful; the FDA does not test a proposed new product but reviews the results of tests arranged by the product's proponent; once a product has been approved, the FDA obtains limited information about bad side effects and even less information about effectiveness. Moreover, neither the FDA nor any other U.S. governmental agency engages in a similar effort to evaluate the safety and effectiveness of new, much less of old, health care procedures and treatments, and to prevent those not shown to be safe and effective from being generally available.

If a government undertakes to prevent the general use in health care of any drug, device, procedure, or treatment whose safety and effectiveness have not been shown beyond reasonable doubt, it will have to assume, pending the review of elements of care already in use, that they are safe and effective if their use conforms to existing standards of practice. However, ongoing review of all the elements of health care technology and appropriate governmental regulation will gradually bring standards of practice into conformity with judgments based on available evidence, gathered and weighed by generally accepted scientific methods. If, moreover, the government allows nothing to be used in any health care it pays for or subsidizes unless using it has been shown beyond reasonable doubt to be both safe and effective, everyone will receive the quality of care that everyone deserves and wants, without being burdened by waste.

Some will object that in undertaking such extensive regulation of health care, a government would subject providers to a large and costly bureaucracy far more responsive to political appointees above it than to the interests of clients who need care by providers they know and trust. In democratic nations, it is of course true that politicians and political appointees exercise ultimate authority over all governmental activities; but the civil servants who regulated the health care industry would have to be competent providers themselves or experts respected by competent providers. Regulatory efforts would succeed only if they had the cooperation of competent providers. Ideally, each professional group regulates itself and disciplines its own members, and that ideal, which is never perfectly realized, ought to shape governmental regulation and be more fully realized by it. With such regulation, providers would be more trustworthy, and people needing care would more likely be able to obtain it from a provider they trusted.

The same government employees who saw to the safety and effectiveness of all elements of the technology used in providing health care could also manage the provision of care covered by insurance and prepayment at government expense.

Thus, not two costly bureaucracies but only one would be needed. The information about the health care each person received, gathered during his or her periodic consultation with a primary care provider, could be the data for ongoing studies of safety and effectiveness. Before being covered at government expense, newly approved elements of health care technology also could be tested more fully for safety and effectiveness by considering the results of using them in the care of those able and willing to pay for them with their own resources.

Those government employees might well also be charged with the tasks of screening the data gathered about the health care people receive and its results, identifying unexplained bad outcomes, systematically investigating them, detecting and deterring deviations from sound practice, and promoting safer alternatives to accepted practices found to lead to avoidable errors.

A century ago, when health care was far less costly than now, avoidable errors by providers and providers' frequent use of elements of technology that were either unsafe or ineffective or both often had tragic results. Today, the tragedy is on a far greater scale, and the colossal waste it involves is intolerable. Governments of wealthy nations could greatly reduce the tragedy and the waste, and they owe it to their citizens to do so. In fulfilling that responsibility, they also will justly benefit poor people all over the world.

Seventh, some health care facilities and individual providers will want to deliver, and some clients will want to receive and be able to pay for, what I shall call "premium care"—elements of health care technology that they regard as better and/or more convenient than "standard care," that is, similar elements of health care technology available at the prices a government sets for whatever is covered by its insurance and prepayment scheme, or partly paid for by its subsidies. The parties interested in premium care cannot be justly forbidden to cooperate in receiving and supplying it, unless that adversely affects standard care or some other aspect of the common good. Of course, in providing premium care, providers ought to conform to just governmental regulations meant to ensure safety and effectiveness.

Some parties interested in premium care will argue that they are entitled to at least part of the governmental payment or subsidy that would go to provide care for them if they settled for standard care. However, giving them even part of that payment or subsidy would promote the proliferation of premium services that would more or less severely limit the availability of standard care and more or less defeat the purpose of fixed prices. Moreover, standard care will meet people's needs, and those who seek premium care can afford it. So, governments justly can, and it seems to me should, refrain from giving those interested in premium care any part of the payment or subsidy enjoyed by those who settle for standard care.

Some health care facilities and individual providers will wish to offer simultaneously both standard care and premium care, as airlines offer coach and first class on the same flights. However, those flying first class subsidize those flying coach, whereas premium care clients would not subsidize standard care clients; and providers offering both sorts very likely would inconvenience standard care clients, at least in scheduling premium care. More important, by billing separately for premium features, providers offering both services could offer premium care at low prices, and thus circumvent the ban on subsidizing it. It therefore seems to me that governments should not cover or subsidize any care whatever supplied by a provider that simultaneously supplies premium care (and by *simultaneously* I mean during a definite period of time such as a calendar year).

Of course, every client should be free to obtain some elements of care from standard providers and others, for which he or she is willing to pay, from premium providers. And standard care providers should remain free to serve a client, such as a foreign tourist, not covered by government schemes, provided they charge no more than the fixed prices for products and services.

Finally, premium care providers should be required to meet the same standard requirements for recording and reporting information about clients, to facilitate enforcement of the regulations meant to promote the safety and effectiveness of procedures and treatments, and the safety of practices. Analysis of the comparative results of premium and standard care also would enable managers to identify superior features of premium care valuable enough to warrant their use in standard care, including technological innovations not yet approved for general availability.

References

Boyle, J.M. Jr. 1977. The concept of health and the right to health care. *Social Thought* 3:5–17.

Boyle, J. 1996. Catholic social justice and health care entitlement packages. *Christian Bioethics* 2:280–292.

Boyle, J. 2001. Fairness in holdings: A natural law account of property and welfare rights. *Social Philosophy and Policy* 18:206–226.

Part VI
Boyle Responds

Chapter 13
An Appreciative Response

Joseph M. Boyle

I should like to begin this response by expressing my gratitude to Christopher Tollefsen and to the authors of the contributions for the attention and respect they show my work by preparing these papers and including them in this collection. They honor me and I thank them. I was surprised by the undertaking when told about it, and I have been regretfully slow to respond.

I was surprised by this tribute because I believe that my work so far is very incomplete and programmatic, even as I come to the end my career as a university teacher. I hope I shall, in whatever years of productive philosophy lie ahead of me in this world, justify the respect and confidence shown by these contributions.

I have been slow to respond not only because I am slow in carrying out philosophical work, which is, for me at least, very difficult, but also because the papers here are so diverse and rich, and raise issues about so many aspects of the philosophical problems I have been concerned with that careful response to the reasoning of these papers must come in pieces and over time. A very short summary of the topics of these papers shows this diversity.

While all the contributions are connected to things I have done, in about half of them, my work is at best a starting point or footnote. Their authors need not have targeted these papers to this collection, and I am very pleased and honored that they have chosen to allow them to be published here. I thank their authors, Don Marquis, Patrick Lee and Robert George, who write on abortion and beginning of life issues, Timothy Chappell, who writes on double effect and doing/allowing, Peter Ryan SJ, who writes theologically about an important argument in the debate over the treatment of those in PVS, Janet Smith, who addresses the right of a competent person to refuse even life saving treatments, and my teacher Germain Grisez, who applies the casuistry of the new natural law theory to current American issues about justice in health care. The other contributions comment more explicitly and closely on my work. Mary Lemmons, a Ph.D. student of mine from the University of St. Thomas in Houston, admirably situates my work and that of the new natural law theorists more

J.M. Boyle (✉)
University of Toronto, Toronto, ON, Canada
e-mail: jboyle@chass.utoronto.ca

C. Tollefsen (ed.), *Bioethics with Liberty and Justice*, Philosophy and Medicine 110,
DOI 10.1007/978-90-481-9791-0_13,

generally. Christian Brugger defends the textual and some conceptual aspects of my work on intention and double effect. Mark Cherry criticizes and Andrew Lustig (mostly) defends my natural law account of welfare rights. Ana Iltis criticizes the new natural law theory and especially my attempt to apply natural law to current debates about global ethics.

In the remainder of this response I will try to put the papers into perspective by showing how these papers are connected to one another and to the relevant parts of my work, and to adumbrate the shape I believe a fuller discussion might take.

13.1 The Substantial Identity Account and Its Implications

Don Marquis has famously argued that abortion is almost always wrong. He has developed a rationale for this conclusion that, as he admits, is quite different from the traditional argument according to which abortion is wrong because fetuses and embryos are human beings. He calls his view "the valuable future view": abortion is wrong for the same reason other killing is wrong; it harms the person killed by depriving him or her of the valuable life they would have otherwise had.

Although Marquis' argument seems sufficient for judging many abortions morally wrong, he also holds that his argument is preferable to the argument that what is wrong with abortion is that it violates rights the fetus has just in virtue of the fact that it is a member of the human species. Marquis accepts that this argument is not speciesist, because membership in the human species is held sufficient for being the subject of rights, and because there is something about being human that establishes humans as rights bearers—rational nature.

Marquis lays out the argument for this view, and criticizes it in detail. This article certainly moves ahead the philosophical debate about abortion. Together with Patrick Lee's response this is an important contribution to which I am very pleased to be associated, if only by way of (fair) critical and (welcome but unjust) appreciative nods to work of mine done long ago. I will highlight just one central issue in this important discussion.

Marquis focuses on a central argument of the substantial identity account. He calls this argument the BLG argument for its authors, Beckwith, Lee and George.

1. All individuals who are rational beings are individuals who have the right to life.
2. All human beings are individuals who are rational beings.
3. Therefore, all human beings are individuals who have the right to life.

Marquis articulates and scrutinizes various possible meanings for 1 and possible arguments for its truth. He concludes that BLG's arguments for its truth are unsound, and that the sense of "rational being" in an interpretation of 1 according to which it is true will render 2 false. But the sense of "rational being" Marquis settles on is, as Lee not surprisingly argues, not what BLG supposes. For Marquis thinks that one can be a member of the human species and still fail to be rational being. BLG believe otherwise.

In arguing that 2 is false, Marquis focuses on anencephalics and those in a permanent vegetative state. They are not rational beings in the relevant sense since they do not have a valuable future, a future that could be experienced and enjoyed. Lee responds that the substantial identity account holds that substantive fulfillment, not simply experienced fulfillment, constitutes value. This exposes an essential point of disagreement—Marquis may have the initially more plausible position to defend, but Lee articulates many of the reasons for the more substantive conception of value. More could be said on this point and no doubt will be. Suffice it to say that Lee's position is a central element of the moral approach he and I share, the so-called new natural law theory.[1] Moreover, the contribution to this volume by Peter Ryan SJ shows in the context of a theological discussion of the treatment of patients in PVS how they can reasonably be thought to benefit from the treatment they receive. The suggestive heuristic of the prospect of eternal life is not, I believe, a necessary condition for thinking that those in PVS benefit from loving care.[2]

Marquis also raises questions about the ability of the substantial identity account of natural human rights to deal with brain death as it is now coming to be understood scientifically. Lee agrees that there might be a living being after the whole brain is permanently non-functioning, but surprisingly he holds out the possibility that such a living thing is not a rational being, since the very possibility of rational activity is lost with the loss of the capacity for sentience. I look forward to Lee's full account and to its further clarification of what he means by "rational being."

Robert George's elegant and informative contribution takes up an issue closely related to the debate between Marquis and Lee, namely, the humanity of the embryo. Along with Lee, George is one of the most prominent defenders of the substantial identity account. Here, after briefly summarizing that account, he applies it to the debate about stem cell research that is destructive of embryos. The main burden of the paper is to show that embryos are human beings: the main claim in support of that is that the development of the embryo from the completion of conception is self-directed. This claim, in turn is based on the scientific evidence. Finally, George refutes the use of scientific and other argument to show that embryos are not living human individuals.

Mary Lemmons' contribution moves out from the focus on abortion and embryos in several directions. The cases she invites us to consider are IVF and euthanasia. The larger themes of her contribution are the role of compassion in thinking about these moral questions and the view of health care in which satisfying patient desire is the fundamental norm, the view she names "instrumentalist bioethics." Lemmons' perspective on this bioethics relies heavily on the jurisprudence which Germain Grisez and I developed in *Life and Death with Liberty and Justice*, thirty years ago (Grisez and Boyle, 1979). She uses our work to show that a competing conception of public morality, one that respects a role for consent by citizens, can shape public morality in accord with the American values of freedom. Whatever qualifications thirty years of dialectic would require Grisez and me to make on the views Lemmons lays out, the basic point she makes seems to me correct: there is a jurisprudence—respectful of liberty and of American values generally—that provides a coherent alternative to the dominant bioethics. With this background in place, Lemmons turns to the question of how we are to understand compassion. Using the new natural law theory, the personalist philosophy of Karol Wojtyla, and the ideas of Victor Frankl, she articulates a sensible account that cuts compassion's connection with instrumentalist bioethics. Within personalism, compassion must be structured and limited by moral standards, and so instrumentalist bioethics wrongly treats compassion for some persons as morally privileged.

Peter Ryan SJ's contribution is a wonderful example of theological refutation. His topic is the position on the treatment of patients in PVS that is most associated with the work of Kevin O'Rourke OP. The view of O'Rourke and his collaborator Benedict Ashley OP is based on a reading of the purposes of human life and human action found in St. Thomas Aquinas' views on the last end of human beings. St. Thomas famously held that all human action has a single ultimate end, and that the true final end of human life is union with God. To simplify a complicated story, one who is permanently unconscious is not capable of pursuing the good, and therefore others have no obligation to keep such a person alive. The inability to pursue the end does not mean that the one so disabled is no longer a person, or that one can kill such an individual, but it does remove any obligation to treat. Ryan focuses on the claims about the ends of human action involved in this account, and explicitly criticizes St. Thomas.

These criticisms, which Germain Grisez pioneered and Ryan has developed over many years, have become an important part of the common approach of the new natural law theorists—one of the few areas where we disagree with St. Thomas as we understand him. Our view is not that there is no ultimate end people should pursue, but that there is no final end so final that its possession makes further motivation impossible. On our account, the final end we should pursue in our choices and other voluntary acts is the divine-human community in which human goods are realized in an integrated way and shared in communion with the divine persons.

St. Thomas' view on the ultimate end plays a limited but important role in O'Rourke's account; it is limited in O'Rourke's argument to decisions about whether or not to treat permanently unconscious patients. This provides a valuable, if not strictly essential, premise in O'Rouke's argument. I say "not strictly essential" because O'Rourke also appeals to the doctrine of ordinary and extraordinary treatments which presupposes neither the Thomistic account of the end of human activity nor his use of the doing/allowing distinction to prevent his reasoning from justifying also killing the permanently unconscious. However, the Thomistic conception of the end of human activity, linked with a strong form of the doing/allowing distinction, is needed if the argument is to proceed in the face the obvious objection, made by Ryan, that the treatments in question are in fact not very burdensome. So, Ryan's critique shows the practical importance of this apparently theoretical disagreement with St. Thomas.

13.2 The Nature and Importance of Double Effect

Christian Brugger's contribution effectively addresses a number of critical responses to my survey of St. Thomas' uses of the expression *praeter intentionem*. The idea that an aspect of an action that is outside the agent's intention is morally "accidental" is important for understanding the doctrine of the double effect. Many people find the source of that doctrine in St. Thomas's treatment of killing in self defense (St. Thomas Aquinas, 1947, II–II, Q. 64, a. 7). There St. Thomas says that the death of the assailant is *praeter intentionem* and therefore accidental to the moral act forbidden by the prohibition against killing by private persons. I think this supposes that bringing about the death is a side effect of a defensive act, which the absolute prohibition of killing by private persons does not automatically exclude. But others, notably Steven Long, disagree, arguing that it is a chosen means. I have not taken up the challenge posed by Long's arguments. Here Brugger does that, and in a way that both improves my account and reveals what is at stake philosophically in these disputes about Thomistic texts.

Brugger's response has a number of significant elements. First, he makes a very compelling textual case for the view that chosen means are in many cases intermediate ends, and since intention is of the end, many means are intended as ends. Second, Brugger acknowledges that one's primary means, what one does without doing anything else to do it, are not as such intended as ends. He then shows that such means are not outside the agent's intention. St. Thomas understands intention as an actualization of the will in respect to the end insofar as the end can be achieved through some means, and choice of the means as based on the intention of the end. So the fact that the primary means is not intended as an end does not show that it is outside the intention, which is the ground for the choice of the means.

Brugger wraps up this textual argument by looking at the context of St. Thomas' treatment of killing in self-defense. That context is the question on the morality of killing in which the issue of self-defense is situated. In the context of the other normative claims St. Thomas makes there—about the permissibility of capital punishment, the wrongness of suicide and of killing innocents—it is hard to see why he would have argued as he did concerning self-defense if his point was that self-defense can justify a choice to kill an assailant.

Brugger then turns to Thomas Cavanaugh's criticism of my view that the inevitability of a side effect's following from one's action is not relevant to whether it is intended—at least in respect to lethal self-defense. Brugger fairly and clearly tracks a necessarily complex argument and improves on my account at certain points.

Finally Brugger takes on the very difficult and vexed idea of the object of the moral act. Here he disputes Steven Long's objection that the physical structure of a chosen act constitutes part of what it is to be a human act. What is thus implicated in an action in that way cannot be a side effect. The distinctive contribution of this closing part of Brugger's paper is its highlighting very effectively the difference between Long's account of the moral object and that developed by Grisez, Finnis and me. Our contrasting view is that in choosing we do a simple physical or mental thing, which is a performance. Insofar as they are events, performances have numerous effects. Some of those effects are understood by us as ways of serving or helping realize some goal we intend. When we choose a performance, we give it an intentional meaning by selecting it to serve a human purpose. Everything else about the performance is a side effect. Such aspects of a performance might or might not be understood, and might or might not be relevant to the overall moral assessment of one's action, but they are not one's end or one's means.

Brugger draws out the practical differences between Long's views and ours by exhibiting how Long's view provides verdicts on some standard bioethical dilemmas that are decidedly different than our account justifies. The difference between Long's own verdicts on two ways to deal with embryos implanted in the fallopian tube seems to me to reveal the implausibility of his insistence that chosen performances have moral significance in virtue of their understood physical properties. He rejects the method of removing the embryo by a drug that breaks down the embryo's attachment to the tube because it also poisons the embryo, but he allows the removal of the part of the tube containing the embryo, presumably because the structure of that action is not death dealing. But in neither case is the choice precisely to end the embryonic life, and in neither case is survival very likely. What if the poisoning effect of the drug took roughly as long to develop as the demise of the child after the removal of the fallopian tube? Long might say this is just question begging: I see no moral difference where he believes one exists. But my point is that the physical structure that imposes intentional meaning on our actions beyond what our intentions require appears arbitrary: why should it be that using the drug is essentially poisoning and removing a fallopian tube containing an embryo that will not survive is not? Both are truly described as killing.

Timothy Chappell's contribution constitutes a significant challenge to my understanding of double effect and of the action/omission distinction. He provides a straightforward and intuitively powerful defense of the ethical significance of the intention/foresight and of the action/omission distinctions. His argument depends on the idea that there are degrees of actionhood. There are paradigm cases of actionhood, that is, cases of intentional action in which persons causally intervene in the world. These cases involve full responsibility. But other cases, in which the central elements in the idea of actionhood are not met or only partially met, involve, typically, diminished responsibility. In cases of omissions there is a low degree of causal intervention; in cases of bringing about side effects voluntarily but not intentionally, there is a low degree of intentional action. So in both, other things being equal, there is less responsibility.

This argument is worked out in careful detail, and I think that it is in those details, not in the initial, intuitively powerful formulation, that grounds for objecting begin to emerge. Here I can only indicate the contours of my alternative approach to these difficult and ethically important matters, in anticipation of face-to-face conversation with the only contributor to this volume whom I do not know personally. That Chappell chose to honor me by including this wonderful piece here is very gratifying indeed.

First, I am a partial reductionist about the moral significance of the action/omission distinction. That is to say, I believe that omissions of one important type are to be handled by way of double effect. They are, in other words, chosen actions like any other chosen actions except that the performance is doing nothing, or not doing something else. Chosen omissions are individuated by the motivational situation in which one faces options; among them sometimes is the option to do nothing, or at least not to do something one finds desirable in some respect. In such cases the fact that one chooses not to act does not seem to me to put the option into a morally special category: one's non-performance is what contributes to one's goal. Understanding causal connections is perhaps necessary for understanding means/ends relationships, but not all of the latter are simply reducible to the means being a cause of the end. Plainly we can realize some ends best by choosing not to do something. I do not see how this fails to be a paradigmatic intentional act. If one chooses not to do something to, let us say, end a human life, then that choice differs only incidentally from choosing to do something for the sake of ending life. Of course, the choice to do nothing that has as a side effect someone's death is not intentional killing. Here the absence of causal intervention seems to me to be morally irrelevant; that the death is a side effect is morally central.

But there is a quite different type of omission that does not involve a choice not to do something, but rather involves a simple failure, without an act of choice at the time, to do what one can and should do. I will call omissions of this type "pure" omissions to distinguish them from cases in which one chooses not to do something (See St. Thomas Aquinas, 1947, I–II, Q. 6, a. 3, concerning whether there can be voluntariness without any act). Obviously, some normative considerations are required to define omissions of this kind. As Chappell notes, without some such considerations, it is impossible to specify among all the things one does not do those non-doings that are failure or omissions. Moreover, Chappell is probably correct in thinking that responsibility for such omissions is diminished. I am uncertain, however, whether the diminution is rooted in the fact that they involve no causal intervention. Perhaps the fact that one has no choice at the time of the omission is more fundamental: the impossibility of doing otherwise does not exist when the thing you should do goes by without a thought. You *could* do it only in the attenuated sense that if you had taken reasonable steps to carry out your responsibility, you would have seen the significance of acting and spontaneously or by choice acted properly. Obviously omissions of this kind are important for moral life, and are at stake in negligence. But the difficulties in assessing the presence of these omissions and any person's responsibility for them are quite different from the casuistical questions surrounding decisions to stop treatment (to end life, respect patient wishes or avoid burdensome treatments) or to refrain from marital intercourse to avoid pregnancy.

In respect to the justification of the doctrine of double effect, our approaches are quite different, and probably incompatible. I think that double effect, or the rule of the side effect as Anscombe called it, restricts the application of moral absolutes to intentional actions. That restriction is necessary since completely general moral norms would otherwise quickly become incoherent. Moreover, the distinction between one's intention in acting and one's voluntarily accepting the side effects of one's intentional actions marks a significant distinction since one can always choose not to perform an intentional action, but sometimes whatever one chooses to do one cannot avoid causing side effects of a kind that it would be absolutely wrong intentionally to bring about.

This view does not suppose that one is less responsible for the known side effects of one's actions than for the intentional actions themselves. In both cases we voluntarily bring them about, ordinarily by choosing to do something. Rather, I hold that different norms apply to side effects than to intentional actions because of differential human power to choose intentional actions and to accept side effects.[3] Chappell, like many others, seems to want to ground the moral significance of the intention/foresight distinction in differential responsibility that I admit has its appeal, as is suggested in the language of the new natural lawyers to the effect that what we choose is self-constituting in a way that accepting side effects cannot be. Perhaps this idea is finally compatible with the rationale for double effect I favor, and therefore perhaps Chappell's elegant argument could fit into a unified account. But I cannot be sure: I began to think about double effect only recently, in 1975.

13.3 Natural Law and Welfare Rights

Mark Cherry addresses my natural law argument for welfare rights. He lays out fully and fairly the arguments I presented in Boyle (2001). In that paper I sought to show that considerations similar to those used by St. Thomas to justify private property also support welfare rights in some social circumstances. The basic ideas in this paper are that the property rights of owners are justified because they secure in an orderly peaceful way a regime for using nonhuman things that facilitates their utility for human benefit. When a group of people have obligations to respond to the needs of others and can do so by cooperative action, I argue that it is unreasonable for them not to so cooperate. So, when politically coordinated action can provide for unmet human need more effectively than can individual action or cooperative action of voluntary societies, those needs should be met by political action, provided that the political action does not deprive citizens of the discretion over their lives which private property protects. The two conditions are important for my account; if political society's capacity to coordinate action to help citizens meet their responsibilities to others cannot achieve that end more effectively than individual or voluntary joint action, then it is not justified; and if the cost of that effort in taxation and other imposition reduces the discretion of citizens over how to live their lives, the effort is unjustified.

Cherry objects to much of this, and articulates many of the reasons, both philosophical and religious, for skepticism about my position. I cannot here respond in detail to these objections. Several points of clarification might move this debate ahead.

First, Cherry notes that I accept a particular view of rights, namely, that articulated by Joseph Raz: a person has a right if and only if another or others have a duty to that person based on his or her need. Cherry lists many conceptions of rights and many ideas about their functions, suggesting that my Razian conception is arbitrary. I accept that I have a particular conception of rights, but I doubt it is arbitrary. Perhaps I could put my point as follows: if a person has a need sufficient to generate a duty on the part of someone or some group to respond to that need, then the person in need has a moral claim that the need be met. If that claim is a claim on a community, then the community has a responsibility to act responsively, and if the community in question is a political society, it must act and do so politically, by taxing fairly and helping fairly and intelligently. Without reference to "rights" the duties of individuals and communities can be delineated. But this analysis is formal; I do not claim to have spelled out the argument from a person's needs to duties on the part of other persons and communities, and I have not indicated the considerations that would make it more reasonable to address obligation-generating human need communally rather than individually.

Cherry notes that the needs of one person do not establish claims without further argumentation. I agree. But the idea that needs of some never generate duties in others, without prior contracts or covenants, seems to me very hard to vindicate. One counter-example Cherry uses is the need of a person to a kidney transplant. I agree that such a person has a need, and that this need does not establish an enforceable right to a transplant. One reason why it does not generate a right is that kidneys are living parts of persons and not exterior things. My argument for welfare rights concerns the use of what Aquinas called "exterior things," sub-human realities, which I believe exclude the human self and its integral living parts. Another reason for thinking this need generates no enforceable claim on people generally is that major operations such as the needed surgery interfere with one's discretion in carrying out the responsibilities one already has. So, such needs as this would not make their way onto my list of welfare entitlements political society could rightly enforce.

Still, I think there could be a duty on the part of some particular person to help that needy person by offering a transplant, and that duty could be based on the donor's recognition of the recipient's interest and of the donor's highly contextualized responsibility to address it. If there are such responsibilities, and if they do not presuppose prior contracts, then however contextualized the process generating the responsibility might be, we have something like a right of one to an action by the other, that is, a moral responsibility that is based on another's need. If Cherry does not think that all such responsibilities presuppose a contract, perhaps we are not far apart at this basic level of what human owe to others.

Cherry rightly notes the difficulty in pluralistic societies of using political authority to provide welfare. This is a real difficulty in making practical the sort of position I have sketched. In my view citizens are abused when required by law to support by taxes actions, such as abortion, which they regard as immoral. However, as with most Catholics who accept the recent Catholic teachings on social justice, I do not believe that rejecting public actions that support such practices as abortion implies that there is no responsibility of political society to help the needy because there is simply no idea of need and appropriate help that could be recognized and fairly provided through the actions of a modern polity.

Cherry closes his contribution by reflecting on the religious significance of charity. There are probably some real disagreements between us here, but not perhaps as wide or deep as he suggests. I agree with Cherry that charity involves a lot more than meeting people's welfare needs. But I do not accept the suggestion that meeting needs has no meaning of its own independent of the moral and spiritual transformation of the charitable on which he focuses: I think charity shows love of God because it serves human need. Meeting those needs is an important part of charity, and this is a goal in which one can be wasteful or effective. When possible without harming other goods or failing in other responsibilities, it seems to me obvious that we should seek to be effective. If the effective and otherwise upright cooperation of people in political society through its actions serves these goals, then one's patriotism and civic loyalty include charitable commitment. Cooperating in political society often involves temptations and compromises, but I am convinced that people can integrate their political activities into their larger religious vocations; the size, the pluralism and the state's ability to use force and coercion do not make political life a Manichean residue of evil that cannot be ordered to God's kingdom.

Of course, there is much in charity that cannot be reduced to providing for the needs that can be met through such things as tax supported services or even voluntary donations of money or goods. The very real limits on what political society can do for the sake of meeting human need points to significant unmet needs not only for things such as food, housing and health care, but for help of various sorts, including reconciliation, evangelizing, inviting others to conversion and praying for them. Individuals' responsibilities for the material welfare of others and for these other needs charity leads us to address are obviously very different. But they are often real and when we carry them out, those charitable actions are part of the gift we offer to God in sacrifice.

Andrew Lustig addresses the same set of issues as Mark Cherry does. He does so, however, from a perspective much closer to my own. Indeed, the pairing of these papers, even if my work were not a target in both, would provide an excellent statement of the issues dividing Catholics who accept Catholic social teaching from Christians of a more libertarian bent. Lustig addresses at some length the worry expressed by Cherry that the use of political society to assist its members in carrying out responsibilities to the poor by establishing welfare rights and taxing them to support it removes the opportunity for exercising charity. Lustig agrees with me that "more than ample opportunities will doubtless remain for the exercise of spiritual and corporal works of mercy, even with a basic welfare system in place." He goes on to observe that the opposed thought is a "curious and unconvincing zero-sum version of virtue theory."

Lustig's paper provides the larger context in Catholic and natural law sources for my argument about property and welfare rights. In doing this he considers a number of important Thomistic texts I do not mention and provides an entry into the current Catholic discussion of welfare rights. This shows that, to the extent my position depends on Aquinas, it can benefit from considerable nuancing. He closes his paper by quoting a ringing statement of the importance of government's providing for the needs of the poor from Pope Leo XIII's great encyclical *Rerum Novarum* (1891). This is an encyclical defending the rights of poor and working people to property, to fair pay and to unite in unions, and it is reasonably thought to be the start of Catholic social teaching.

Lustig reviews in detail the Thomistic arguments about property that are the basis of my position on welfare rights. He is concerned with the requirement of common use, what is now called in Catholic teaching "the universal destination of material goods." This is the idea that the goods of the world are not for the use of any individual person but are created by God for the common benefit. I understand this to mean that God's sub-human creation is all there for human benefit. Private property involves dividing this common store of goods up into goods that belong to some and not to others. This division is justified pragmatically; it avoids conflicts over things and increases the likelihood that they will be well used. But private ownership does not cancel the idea of common use. Indeed, the discretion over property by owners does not mean that its utility for benefitting humans more generally can be ignored. The owner is in authority but directed by moral consideration to use his goods well and fairly.

I believe that Lustig and I have a disagreement about how exactly this is to be understood, but I am unsure how deep or important our disagreement is. If I understand him, he believes that it is important to be able to say that, before its division into private property by political society, goods are held in common; they belong to everybody. I have resisted that formulation because it implies that one regime of property is replaced by another. If it makes sense to say that something is owned by everybody, then a person's appropriating it seems to be taking as his own what belongs to others. My simpler story is that common use refers to the fact that owners have a responsibility to see to it that the parts of the world under their authority are used for human good, and not simply their own good. The world before private property is not owned by anyone, but was created for human use (and appropriating parts of it serve that use, if the benefits are fairly distributed).

13.4 Natural Law in Theory and Practice

Ana Iltis' contribution responds to my efforts to address issues of global bioethics, in particular, the aspiration for common moral standards, from the perspective of natural law theory. She briefly and fairly summarizes my account of the features of natural law moral theory that render it appealing in this area: its rationalism, its universalist nature, and its account of moral error.

She responds to this account by disputing at length the defense of these features of natural law in the new natural law theory. There is much in Iltis' objections to natural law that calls out for a response I cannot provide here. I would like to take note of one theme of her rebuttal that seems to me especially significant. Iltis notes several times that the natural law verdict on moral issues is rejected by many people, including other moral rationalists and universalists. This suggests that there is no neutral perspective above the fray of competing ethical loyalties.

To clarify where I stand on this, I should note that, like most defenders of natural law, I am acutely aware of the phenomenon of radical moral disagreement. Moreover, I do not believe that any ethical theory is above the fray of the dialectic of arguments about moral theory and its implications. If natural law theory is correct, there will be principles all can know, but any attempt to indicate and defend them are by people with a limited and incomplete perspective. As for the extent of moral disagreement, I believe that natural law theorists are required to provide an account of it. I have used St. Thomas' discussion of moral knowledge and error to point to the main elements of that account (Boyle, 2004a, 2006, 2010). But it remains to be developed.

After criticizing these aspects of natural law theory, Iltis turns to a critique of my claims about the extent to which there are commonalities in moral judgment across communities and those with different moral theories and ideologies. My formulation of the existence of such commonalities owes much to Michael Walzer's conception of a common moral world in *Just and Unjust Wars*. He too recognizes that moral theories differ and that in many particular moral questions differences are resistant to resolution. Still, people do argue about moral matters and there is enough consensus for the argument to proceed and sometimes to reach an agreed upon conclusion. The moral presuppositions for that seem to include some fairly general and somewhat thin moral beliefs that are shared. I have suggested that such beliefs would be a kind of modern *jus gentium*, that is, a part of positive morality that is widely accepted. I have asked whether any widely shared moral beliefs, especially in bioethics, might qualify. I suggested that the right of competent persons to refuse medical treatment is one such belief. If there are such beliefs, natural law can account for their widespread acceptability. Iltis argues that most of this is simply false. Her evidence for this claim is not easily rebutted, but it is important to note that my claims about moral agreement are abut *thin* moral agreement, and the fact that many who hold the same moral beliefs do so for different reasons is acknowledged in the natural law account of such agreement as there is.

Iltis then turns to a critique of the thinness of my conception of shared norms. She zeros in on the issue of the vaguenesss of moral terms in a thin conception of moral norms. Iltis develops this with a very interesting case, Angeles tan Alora's account of the understanding of honesty in Filipino culture. The value of honesty is defined in personalist, not legalist terms, as a matter of the advantages to the person, his or her family and group, not to society. This allows that stealing from one with whom one has a close relationship is wrong but an honest person may steal from government. This is a very interesting case study. But it is unclear from Alora's account that natural law would have nothing to say about such modes of moral thinking. Is the limitation of the scope of honesty defensible? Why should the reasons for local honesty not apply further than one's group? Vagueness about what honesty amounts to can surely be reduced by honest discussion.

Finally, Iltis addresses the use of state authority to enforce a universal morality. As Cherry's and Lustig's contributions indicate, I believe that some welfare rights are enforceable by political society; but, like most people, I also believe that regulations of such things as the practice of medicine and its research within a polity can be morally justified. However, I have argued at length that there is no global bioethical authority with positive authority, and so I believe that any enforcement of bioethical regulation that has a global dimension must be by the political leaders of the various polities, including their cooperation with other nations through treaties. I do not see how that view is guilty of a charge of using political power to enforce controversial universalist ethics. The norms of natural law that are relevant to this issue are those that limit political authority to the service of the common good of a community. These norms and the fact that there is no universal global community limit what trans-national authorities can rightly do.

Janet Smith is a self-described as a proponent of natural law. Her approach is in its detail quite different than mine, and so provides a usefully different counter-point to Iltis' rejection of natural law. That is so even though Smith's reflection on the state of bioethics includes considerable skepticism about bioethicists' interest in rooting bioethical judgments in any moral theory, natural law included. Smith's contribution deals with the role of the physician in medical decision making, and her concerns are framed by an acceptance of the right of competent persons to refuse medical treatment. Smith's main question is whether we should understand that widely accepted right as severely limiting physicians' persuasive role in medical decision making. Among the dominant concerns in this contribution are several that tie it to other contributions. Smith shares Mary Lemmons' concern that medicine should not and does not simply serve patient desires, and although she states the Christian case in favor of autonomy, she is skeptical that it is an absolute value and also skeptical that anyone seriously thinks so. Finally, Smith reminds us of Aristotle's emphasis on the importance in both medicine and morality of concern for the particular circumstances and context of decisions.

But that concern for particulars is not on Smith's view a bar to a natural law understanding of how people with very different moral outlooks can and sometimes do come to agreement. Smith provides an actual example of the working of an ethics committee in the face of a difficult situation in which a pregnant woman refuses antibiotics that are needed to treat sepsis in her uterus. Her reasons for refusal were ungrounded, but firmly held, views about her own sensitivity to the antibiotics. Her will in the matter was simply not respected by the health care team or the committee, at least not without much push back. Various efforts were made by the committee (composed of many who did not accept full human standing for her fetus) to persuade her and otherwise to block her decision on the grounds that her disregard for the fetus was over the line. Finally the problem was resolved by persuading her homeopathic doctor, on whom she relied for her views about her sensitivity to the antibiotics, to talk her into taking the medicine. The established positive morality favored going along, but there were grounds for critique and for an emergent agreement about the case among those with differing moral outlooks. Smith's explanation is that the most basic moral considerations of the natural law were in play in these conversations, and that they favored *guided* autonomy, not simply autonomy as whatever the patient decrees. The vagueness of the boundaries of a norm that has become established in positive morality can be reduced by such considerations—considerations many can agree are relevant even as they differ on other moral issues. Cases like this seem to me as important for thinking about principles and bioethics as is Iltis' report of Alora's story about honesty.

I suspect that Smith may think it easier to set aside the right of a competent person to refuse medical care than I do. However, that possible disagreement takes place within a background of agreement about the mutual responsibilities of people cooperating in common action, namely, the responsibility to communicate about the morality and wisdom of actions in which they are both involved. That responsibility is not eliminated by the fact that one party clearly has the authority to settle what will be done in a particular case.[4] So, moral persuasion is appropriate in the physician/patient relationship. Once this is accepted the issue becomes what forms of influence are appropriate in fulfilling this responsibility, and Smith explores this in a rigorous way. I understand the right to refuse medical treatment as the specification of the final authority in medical decisions as belonging to the person involved. Smith does not say exactly this, but neither does she deny it nor do her arguments contradict it.

Germain Grisez's contribution, which addresses the current moral difficulties in fairly paying for American health care, shares Smith's concern that all the moral issues that emerge in such social problems must be addressed, and that ethical solutions cannot solve some problems, such as costs and health care for the poor, while ignoring others, such as the waste inducing moral hazard generated by most third payer forms of insurance. Grisez begins with an account of the dynamism of modern health care and indeterminacy but urgency of people's reasonable need for it. He indicates reasons why health care cannot be provided through markets as other commodities are, and the factors that make unfeasible its provision by charitable institutions to those who cannot pay for it.

Grisez argues that in schemes such as insurance or prepayment that cover a broad variety and unspecified quantity of health care services costs tend to outrun the willingness to pay of those who must. Efforts to deal with this tendency are almost all deeply flawed. Some are unfair for the users or classes of users; some to the payers; some also to the providers, who are asked to arbitrarily cut service or to manage cases they cannot understand. Grisez goes on to list some special moral flaws of American health care delivery, starting with the often criticized policy of allowing heath care insurance costs to be paid with pre-tax dollars, an obvious injustice to those not eligible for such plans, and including the notorious gaps in Medicare and Medicaid. Grisez then turns to proposing seven principles that might remedy this morally unsatisfactory state of affairs. He argues first that government should pay for the provision of health care to the poor at the level it pays for the health care of its own employees and dependents. Second, governments should set prices for paying providers of health care to the poor, and that payment will not be fair if it is the minimum such providers would accept. Third, health care services that can be insured should be limited so as to remove as much as possible the moral hazard and ensuing temptations and waste created by blanket coverage of health care needs. Fourth, the government should be responsible for paying for the insurance for the insurable parts of the health care of the needy. Moreover, it is not unreasonable, and some moral problems are justly avoided, if the governments with the capacity pay this insurance as a public benefit for all. Fifth, clients should as much as possible pay for their uninsured health care; the poor should be helped in this and in other areas of need by subsidies they can use as they see fit. Sixth, regulation of health care technology and drugs, not only for safety but for effectiveness, is a grave governmental responsibility. Seventh, premium care cannot be fairly prohibited, but should not be subsidized or allowed to compromise standard care.

In offering this analysis, Grisez is not naïve about the fact that such conditions are not likely to me met in the policies politicians and health providers are likely to make. The exercise is in rational casuistry, not lawmaking, and the aim is to articulate the reasons that could help form the consciences of those who want to design or support just policy and are willing to engage with the analysis he presents.

This contribution, from my Ph.D supervisor and the initiator of the movement others have called the "new natural law theory," provides an appropriate closing for this volume. It shows our commitment to applying moral principles carefully to fully specified actions and policies. The strategies on display here, especially Grisez's use of the Golden Rule, are the common property of those in the natural law tradition. The rigor of the reasoning on complex topics, to illuminate the issues and choices people face, is a great example of the moral undertaking in which Grisez and I, along with John Finnis, Robert George, Patrick Lee, Peter Ryan SJ, Christian Brugger, Christopher Tollefsen and others have been engaged these last four decades.

Notes

1. See Finnis et al. (1987, 303–309).
2. See Boyle (1992, 1995).
3. See Finnis et al. (1987, p. 292) for the first statement of this view: for my most recent effort on this see Boyle (2004b).
4. See Boyle (1981).

References

Boyle, J. 1981. The patient/physician relationship. In *Moral responsibility in prolonging life decisions*, eds. D. McCarthy and A. Moraczewski, O.P., 80–94. St. Louis, MO: Pope John Center.

Boyle, J. 1992. The American debate about artificial nutrition and hydration. In *The dependent elderly: Autonomy, justice and quality of care*, ed. L. Gormally, 28–46. Cambridge, MA: Cambridge University Press.

Boyle, J. 1995. A case for sometimes feeding patients in persistent vegetative state. In *Examining euthanasia: Legal, ethical and clinical perspectives*, ed. J. Keown, 189–198. Cambridge, MA; Cambridge University Press.

Boyle, J. 2001. Fairness in holdings: A natural law account of property and welfare rights. *Social Philosophy and Policy* 18:206–226.

Boyle, J. 2004a. Natural law and global ethics. In *Natural law and the possibility of a global ethics*, ed. M.J. Cherry, 1–15. Dordrecht, The Netherlands: Kluwer.

Boyle, J. 2004b. "Medical ethics and double effect: The case of terminal sedation. *Theoretical Medicine and Bioethics* 25:51–60.

Boyle, J. 2006. The bioethics of global biomedicine: A natural law reflection. In *Global bioethics: The collapse of consensus*, ed. H.T. Engelhardt, Jr., 300–334. Salem, MA: M&M Scrivener Press.

Boyle, J. 2010. The ethical significance of moral disagreement. In *At the roots of Christian bioethics: Critical essays on the thought of H. Tristram Engelhardt, Jr.*, eds. A. Iltis and M.J. Cherry, 229–246. Salem, MA: Scrivener.

Finnis, J., Boyle, J., and G. Grisez 1987. *Nuclear deterrence, realism and morality*. Oxford: Oxford University Press.

Grisez, G., and J. Boyle 1979. *Life and death with liberty and justice: A contribution to the euthanasia debate*. Notre Dame and London: University of Notre Dame Press.

St. Thomas Aquinas. 1947. *Summa theologiae*. New York, NY: Benzinger Brothers.

Index

C. Tollefsen (ed.), *Bioethics with Liberty and Justice*, Philosophy and Medicine 110,
DOI 10.1007/978-90-481-9791-0, © Springer Science+Business Media B.V. 2011

FSC
www.fsc.org
MIX
Papier | Fördert
gute Waldnutzung
FSC® C083411

Kolibri 360 GmbH
Ferdinand-Jühlke-Straße 7
99095 Erfurt, Deutschland
produktsicherheit@kolibri360.de